Y0-BUP-300

Reflections on Symmetry

Distribution
VCH, P.O. Box 101161, D–6940 Weinheim (Federal Republic of Germany)
Switzerland: VCH, P.O. Box, CH–4020 Basel (Switzerland)
United Kingdom and Ireland: VCH (UK) Ltd., 8 Wellington Court,
Wellington Street, Cambridge CB1 1HZ (England)
USA and Canada: VCH, 220 East 23rd Street, New York,
NY 10010-4606 (USA)

ISBN 3-906390-01-2 (VHCA, Basel) ISBN 3-527-28488-5 (VCH, Weinheim)
ISBN 1-56081-254-0 (VCH, New York)

Edgar Heilbronner
Jack D. Dunitz

Reflections on Symmetry

in Chemistry.... and Elsewhere

Illustrations by
Ruth Pfalzberger

Verlag Helvetica Chimica Acta, Basel

Weinheim · New York · Basel · Cambridge

Prof. Dr. Edgar Heilbronner
Grütstrasse 10
CH–8704 Herrliberg, Switzerland

Prof. Dr. Jack D. Dunitz
Obere Heslibachstrasse 77
CH–8700 Küsnacht, Switzerland

Published jointly by
VHCA, Verlag Helvetica Chimica Acta, Basel (Switzerland)
VCH Verlagsgesellschaft mbH, Weinheim (Federal Republic of Germany)
VCH Publishers, Inc., New York, NY (USA)

Editorial Director: Dr. M. Volkan Kısakürek
Production Managers: Jakob Schüpfer
Layout and Design: Yann R. Keller
Cover Design: Ruth Pfalzberger, Bruckmann & Partner, Basel

Library of Congress Card No, applied for.

A CIP catalogue record for this book is available from the British Library.

CIP-Titelaufnahme der Deutschen Bibliothek

Heilbronner, Edgar
Dunitz, Jack
Reflections on symmetry in chemistry and elsewhere / Edgar Heilbronner, Jack Dunitz. –
Basel: VHCA; Weinheim; Basel; Cambridge;
New York: VCH, 1993
ISBN 3-906390-01-2 (VHCA, Basel)
ISBN 3-527-28488-5 (VCH, Weinheim)
ISBN 1-56081-254-0 (VCH, New York)

Printed on acid-free paper.

Printing: Birkhäuser+GBC AG, CH–4153 Reinach BL
Printed in Switzerland

PREFACE

This book has its origin in a lecture delivered by E. H. at the Erni Museum in Lucerne in 1980. It was the first in the annual lecture series 'Panta Rhei' (under the auspices of the Hans-Erni-Stiftung), dealing with the border areas between art and science, and aimed at a non-specialist audience. An expanded version of this lecture was published by the Hans-Erni-Stiftung, Verkehrshaus Luzern, in 1981 in a limited edition – a book now out of print – under the title *Über die Symmetrie in der Chemie*. An abridged version of the same lecture, given before the Akademie der Wissenschaften in Göttingen, was published in the *Jahrbuch der Akademie* in 1986.

The present volume is an English language augmentation of the earlier book. Over the years several of its readers, particularly Vladimir Prelog, Sason Shaik, and J. D. D., had urged that it should be available to a wider circle. In the end, J. D. D. took on the task of translating the text and, as things turned out, several sections were expanded in the process, and new topics added. However, it would probably never have come into existence without the persistence of our friend and colleague M. Volkan Kısakürek, who has acted as a catalyst through his constant enthusiasm, encouragement, and drive. He is responsible for the layout and production of this book. We are also much indebted to Ruth Pfalzberger for her friendly collaboration in producing the graphic artwork.

Edgar Heilbronner
Jack D. Dunitz

For there is a music wherever there is a harmony, order or proportion; and thus far we may maintain the music of the spheres; for these well ordered motions, and regular paces, though they give no sound unto the ear, yet to the understanding they strike a note most full of harmony.

Sir Thomas Browne (1605–1682)
Religio Medici

I.

Originally goddesses of memory only, the Muses are credited with the encouragement of the creative arts and sciences – after all, one of the nine, Urania, was entrusted with responsibility for astronomy, the only natural science pursued in antiquity. Yet the picture of the scientist depending on inspiration from the Muses may strike most people as somewhat incongruous. They would acknowledge that scientists are creative in a certain way but would view this kind of creativity as different in essence from the creativity of poets or painters. Perhaps the difference is not so great as one might imagine. Artists and scientists are both driven by a kind of fascination with the mystery of human experience and the search to give meaning to it. Moreover, although most people may think of scientific thought as a purely deductive process and of scientists as basically serious-minded, fantasy is an essential ingredient of creative science, and sometimes even a dash of craziness is needed when truly original ideas are called for. Of course, science is an inescapable feature of our modern world. For some it has a good image, for others a bad one, and in either case it has to be taken seriously, especially with regard to its unpredictable consequences. Likewise, for its practitioners it may be a calling or even a religion, for some it is an obsession, for others it is big business. But among these many aspects there is also in every scientific activity an element of play – the kind of play indulged in by children, which can be deadly serious or frivolous or, more often, both simultaneously. Perhaps one should include this playful quality in the activities encouraged by the Muses.

In this book we shall first try to show how such a primarily 'playful' approach to symmetry has influenced the early development of one branch of the exact sciences, namely chemistry. We then take a few examples from more modern developments to show how formal symmetry considerations have become fundamental for a deeper understanding of molecular structure and reactivity.

One difficulty we have to face is that the interconnection between the playful aspect of symmetry on the one hand and an exact science on the other hand is not at all simple. It has to do with that jumble of formulas that may well evoke painful memories of tedious school lessons. This makes it hard to steer a secure course between the Scylla of irresponsible superficiality and the Charybdis of unintelligible jargon. A treatment rigorous enough to satisfy finicky criticism of experts and at the same time gentle enough for the non-specialist is just not possible in the limited space available. At best we can strive for an impressionistic picture, similar to Monet's painting of Rouen Cathedral, which, closely viewed, is seen to lack all exact detail. It is intended to convey merely an overall impression of the building, of the interplay of light and shadow, even at the risk that the paint may be applied here and there somewhat too thickly and so cover or even falsify details which are irrelevant for us but may be precious to the expert in medieval architecture.

II.

The world is chiral and clinal,
enjoy symmetry wherever you find it.

Vladimir Prelog

The world around us is in general so little symmetrical that our recognition and awareness of striking exceptions may become lasting experiences. Fortuitous symmetrization, such as reflection in a calm mountain lake, can surprise and please us; on the other hand, its pictorial representation can all too easily acquire a kind of artificial quality, sometimes even an element of kitsch – although this cannot be said to apply to Hodler's painting of the Thunersee reproduced below.

Thunersee by Ferdinand Hodler[1].

A few examples of symmetry in Nature.

On the other hand, the pattern of stars in the Firmament lacks any symmetry in spite of the elegant elliptical orbits traced out by the planets around the sun in the course of time. Thus, the need to organize groups of stars together in separate compact though unsymmetrical constellations, as a kind of mnemonic device, had been recognized in very early times as an aid to finding one's way in the unsymmetrical glitter of the night sky. In this low-symmetry environment, the symmetry of small objects, in animate and inanimate Nature, as revealed in flowers, leaves, animals, and in crystals, must have struck early man from his very beginnings as conspicuous.

Such symmetry could have been perceived as 'beautiful' and even – as we shall see later – viewed as the expression of some deeper lying order. It is, therefore, not too surprising that in the gradual course of human development, people set about enhancing the beauty of all sorts of objects with little or no symmetry by artificially 'raising' their symmetry. For example, by polishing, a crystal could be shaped into a gemstone with regularly arranged facets and thereby made more attractive – and more valuable.

A few examples of gemstones.

Symmetrization or symmetry augmentation has been used from earliest times as a means of producing 'beauty'. Periodic repetition of arbitrary, unsymmetrical elements in one dimension can be combined with reflection to produce band patterns that can be followed through all cultures, beginning with the wood-carvings of primitive peoples, through the friezes of classical temples, right up to the silk ribbons still traditionally produced in Basel today.

Band patterns, from classical friezes to silk ribbons.

If the symmetrization process is extended into the plane, that is to say, by periodic repetition in two dimensions, combined with reflection and rotation of the element of pattern, one can produce the familiar diversity of tilings, of carpet and curtain patterns, of parquet floor patterns, and of Christmas present wrapping papers, in inexhaustible variety. Many of us know and admire the highly original and sometimes bewildering examples by the Dutch artist Maurits Escher, who occupied himself deeply with the problems of periodically symmetric patterns. Less well known are the earlier examples by Koloman Moser.

Symmetric, periodic patterns by Koloman Moser (top) and Maurits C. Escher[2].

Wall decorations in the Sala del Reposo, Alhambra.

Escher was inspired towards this unconventional style by a visit to the Alhambra palace in Granada, one of the wonders of the world as far as decorative symmetry is concerned. The artistic drive of the high Moorish civilization in Spain was restricted to the exploitation of abstract colored patterns, visual imagery being forbidden in the Islamic faith. Not being limited in this way, Escher developed his own strange representational art, using fish, horses, lizards, human figures, etc., as the basic units of pattern, although, as he himself said, these figures were not consciously planned but were forced on him by the strict rules of periodic space-filling.

The feeling for symmetry seems to be inborn in mankind. At a not particularly high level, it can lead to a lemming-like urge – well known to art dealers among their customers –, a craving for a mirror-symmetrical complement to every object that passes for a work of art: for every little angel on the left, a little angel on the right must be found, in order to satisfy the symmetry-determined aesthetic. This could be summarized in pseudo-mathematical form by the equation:

A pair of (almost) symmetric putti.

$$\text{beauty} = \text{constant} \times \text{symmetry}$$

The validity of this equation can hardly be denied by anyone who, even as a child, has experienced the gratifying fascination of a kaleidoscope.

It is hardly surprising that the idea gradually arose that the forces which hold the world together must satisfy the desire for symmetry, even though this underlying principle might not be perceptible at first sight – or even at second sight – in the macroscopic material objects accessible to our senses. In relation to Nature, one found oneself in a somewhat analogous situation to those critics who want to ascribe the perfect balance of Raphael's Madonna alba, for example, to the artist's conscious awareness of an underlying pentagonal symmetry

Madonna alba, by Raphael[3].

In other words: The Gods must share with us this special liking for symmetry, even when it may not always be perceptible on the surface and – as we shall see later – must sometimes give way to still more important principles.

III.

Der Regen ist eine primöse Zersetzung luftähnlicher Mibrollen und Vibromen, deren Ursache bis heute noch nicht stixiert wurde.

Karl Valentin

The idea that the material world around us is composed of very small particles was put forward by Leucippus of Miletus (ca. 475 B.C.), a member of the Pythagorean school, and it was then taken up by Democritus of Abdera (ca. 460–370 B.C.). He taught that these particles, the so-called atoms, are indivisible (hence the name) and in constant motion, and that their impalpable but nevertheless quite definite shapes determine the properties of substances. So, for example, the atoms of gases and liquids were supposed to be spherical and able to glide smoothly over one another, while those of solids were hard and rough and interlocked. He was also of the opinion that the taste and odor of substances were determined by the shapes of their constituent atoms. To be sure, this was no physical scientific theory in the modern sense, nor were any conclusions drawn that might have been experimentally verified. Two thousand years ago, one was still far from the recognition of the importance that planned experimentation would come to occupy in modern science – for this step we are indebted in the first place to Galileo.

The classical four elements, fire, air, water, and earth were proposed by Empedocles (ca. 490–430 B.C.), who was also the first to write something vaguely resembling a chemical formula. For example,

blood = 1/4 fire + 1/4 water + 1/4 earth + 1/4 air
bone = 1/2 fire + 1/4 water + 1/4 earth.

These four elements were then taken over by Plato (427–347 B.C.), who rejected the atomic hypothesis of his contemporary Democritus. Plato, as a Pythagorean, associated one of the first four regular polyhedra (the so-called platonic solids) with each element: tetrahedron = fire; octahedron = air; cube = earth; icosahedron = water.

Platonic solids as symbols for the four elements and the quintessence.

These symbols had much the same significance as pre-1800 chemical formulas, that is to say, before the emergence of the Daltonian atomic theory. Symmetry played an important role in Plato's considerations. From the fact that the symbols – tetrahedron, octahedron, and icosahedron – could be decomposed into equilateral triangles which could be re-assembled to form the other polyhedra, Plato drew the conclusion that fire, air, and water could be transformed into one another. On the other hand, since it was not possible to decompose the cube, the symbol for the element earth, into equilateral triangles, but only into squares, he concluded that earth could not be transformed into fire, air, or water.

The dodecahedron, bounded by regular pentagonal faces, was added later, as the fifth essence – the quintessence – reserved as symbol for the Universe, presumably because the construction of the pentagon with circle and straight-edge was at that time considered to be a triumph of mathematics. The quintessence, of a purer quality than the other four elements, was supposed to have been formed

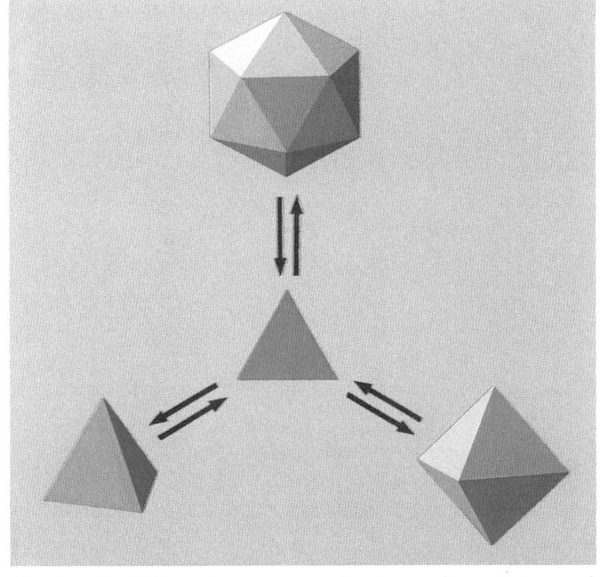

A classical Greek view of the transformations of fire, air, and water into each other.

at the creation of the Universe, when it flew upwards to form the stars. The pentagon and the related pentacle or five-pointed star seem to have retained a special kind of magical significance as symbol for purity well into modern times. They were – and possibly still are – used to help to ward off evil spirits and witches. "Das Pentagramma macht dir Pein?" asks Faust to Mephistopheles.

From the purely scientific point of view it is perhaps not so important that Plato proposed these highly symmetrical polyhedra as symbols for the elements. Nevertheless, it is remarkable that those early philosophers were intuitively convinced that the fundamental building blocks of the elements, or at least their symbols, could not have any arbitrary potato-like shapes but must have a pleasing, highly symmetrical form.

A special kind of symmetry consideration that goes back to Aristotle (384–322 B.C) depends on the relationships among the four elements and the properties that characterize them. If these are represented graphically, one can produce a symmetrical scheme which, allowing for some exaggeration, can be described as the 'periodic system' of the classical four elements. If one of the four had been unknown, its existence could have been predicted from the position of the gap in relation to the remaining three elements, plus a good admixture of fantasy. In spite of their far-reaching idealistic content, these ideas of Democritus and Plato have continued to be effective and are the acknowledged precursors of modern atomic theory.

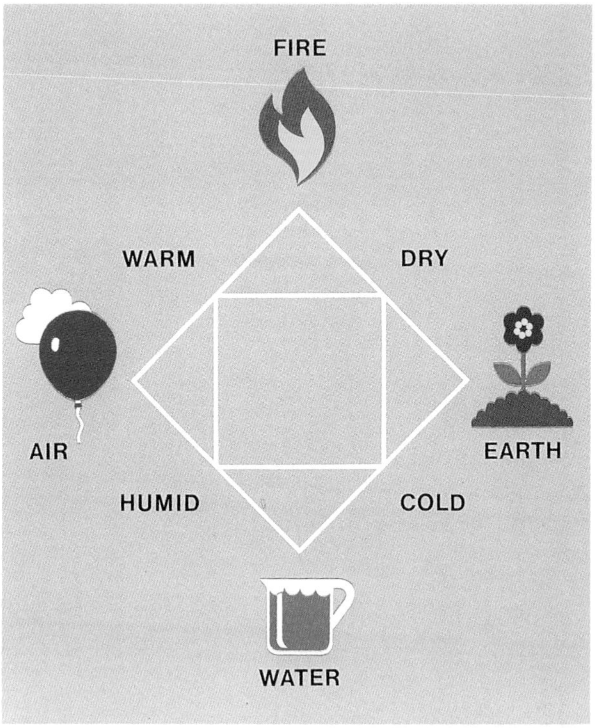

The classical Greek 'periodic system' of the elements.

Since we are here not primarily concerned with in the history of chemistry, we jump to the 17th century, more precisely to the year 1611, when Johannes Kepler (1571–1630) presented his patron Johannes Mathäus Wacker von Wackenfels as a New Year's gift a treatise on the subject of hexagonal snowflakes (*De Nive Sexangula*)[4]. In this work Kepler tried to explain the macroscopic symmetry of a crystal by examining the manner in which the elementary building blocks might pack together. The specific example he chose was the hexagonal symmetry of snowflakes. In this, of course, he did not think in terms of anything resembling what we would now call molecules, but rather in terms of spherical particles of water or ice. Only later, with the emergence of atomic theory, and in particular after the work of Dalton, did it become clear that molecules are in general not spherical but have more or less definite, usually unsymmetrical, shapes. We shall, however, disregard this historically important fact and take up Kepler's problem, namely the answer to the question of which macroscopic symmetries can result from the packing of spheres. It is worth mentioning that the same problem had been successfully tackled by Thomas Harriot (1560–1621)[5] a few years earlier, in 1599. But, as it often happens in science: Kepler published first!

IOANNIS KE-
PLERI S.C. MAIEST.
MATHEMATICI
STRENA
Seu
De Niue Sexangula.

Cum Priuilegio S. Cæs. Maiest. ad annos xv.

FRANCOFVRTI AD MOENVM,
apud Godefridum Tampach.

Anno M. DC. XI.

Title page of Kepler's De Nive Sexangula[4].

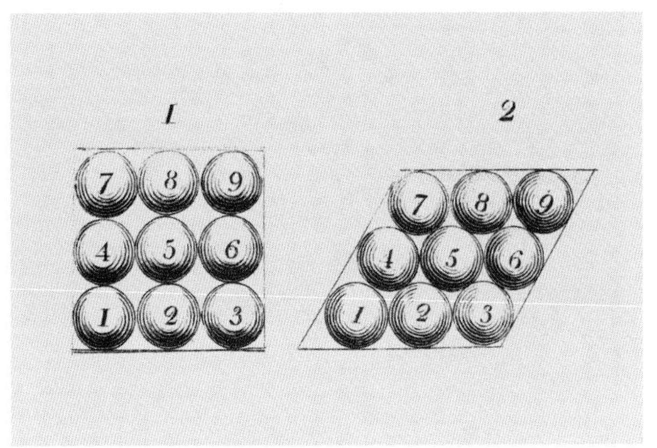

Square and hexagonal arrangement of touching packed spheres (Dalton)[6].

If one covers a two-dimensional flat surface with identical spheres, as illustrated for spherical atoms in the picture taken from John Daltons's book, *A New System of Chemical Philosophy*, published in 1808, and demands that each sphere is in contact with other spheres, there are just two highly symmetrical arrangements.

In arrangement (1) each sphere is in contact with four others (for example, sphere 5 touches spheres 2, 4, 6, and 8), leading to a square lattice; in the other arrangement (2) each sphere has six nearest neighbours (for example, sphere 5 touches spheres 2, 3, 4, 6, 7, and 8), with the result that a hexagonal lattice is formed with acute angle of 60° and obtuse angle of 120°. In the latter arrangement the density of spheres on the surface is maximal. Starting with this densest packing of spheres in the plane and adding further spheres according to the same contact pattern, one obtains shapes (as shown in the illustration taken from the same book) that make it evident that this model leads to figures, or at least can lead to figures, with the typical hexagonal habit of snowflakes.

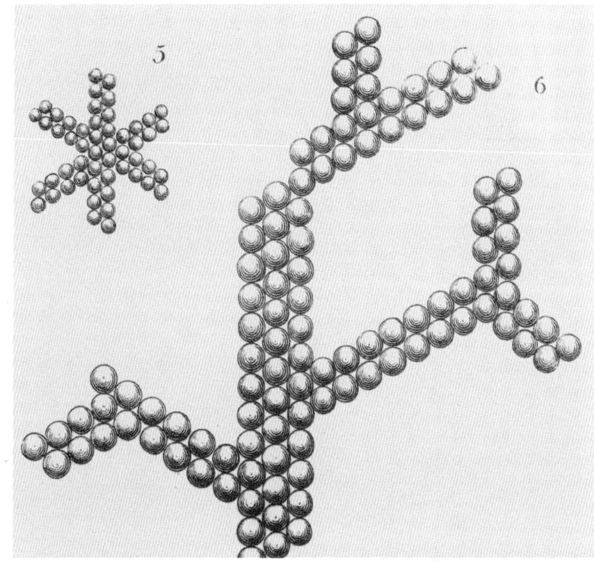

Extended planar patterns made from hexagonally packed spheres (Dalton)[6].

One page of snowflakes from Snow Crystals *by Bentley and Humphreys*[7].

The diversity and beauty of snowflakes, which occur in an infinite variety of shapes despite their selfsame symmetry, has been captured in the classical collection of micro-photographs made by W. A. Bentley and W. J. Humphreys, published in 1936, a sample of which is shown.

When one examines Kepler's theory of snowflake formation more closely, one notices that although it leads to local hexagonal symmetry it does not account for the striking regularity of the branching in the snowflake as a whole, as was indeed mentioned by Kepler himself in his treatise. In spite of this shortcoming, which has not been satisfactorily resolved up to the present day, the model represents a very important step towards the explanation of the macroscopic symmetry properties of matter as a direct consequence of its invisible, sub-microscopic structure. This becomes especially clear when we consider briefly, instead of the two-dimensional case, the three-dimensional one, which was also already investigated in the early studies of Harriot and Kepler.

When we leave the planar arrangements of mutually touching spheres and examine how they can be made to fit together under the same conditions in space, we find that a given sphere can be in direct contact with a maximum of exactly twelve other spheres. There are two ways of accomplishing this. In one case (left) the upper triplet of spheres projects directly over the lower triplet, while in the second case (right) the upper triplet is rotated by 60° with respect to the lower one.

The two arrangements of twelve spheres in contact with a central one.

If one continues these two arrangements in a regular way, one obtains two kinds of densest packing of spheres in space. The first (left) has hexagonal symmetry, the second (right) has cubic symmetry, as can be most easily seen with the help of models. Although the densest packing of circles in the plane has been rigorosly proved, this has not yet been accomplished for the three-dimensional case if one allows the possibility that some irregular, i.e., non-periodic, packings may have a higher density than the periodic one. However, no one seriously doubts that the hexagonal and cubic arrangements do represent the densest three-dimensional packings of spheres. (At time of writing, the Chinese mathematician Wu-Yi Hsiang claims to have established just such a proof[8].)

Kepler tried to explain not only the shapes of snowflakes but also the shapes of honeycombs and of pomegranate seeds. The honeycomb problem is similar to the snowflake one, both depending on arrangements of hexagonal cells in the plane; the pomegranate problem is a little more complicated, and in the course of solving it Kepler made a major discovery. He found that just as it is possible to cover the plane with parallelograms, equilateral triangles, squares, and hexagons, it is possible to fill space not only with parallelepipeds, triangular, square, and hexagonal prisms but also with a special type of polyhedron, the rhombic dodecahedron, a convex figure with 12 faces and 14 vertices – which can be considered as the 8 vertices of a cube plus 6 additional ones above

Hexagonal and cubic close packings of spheres.

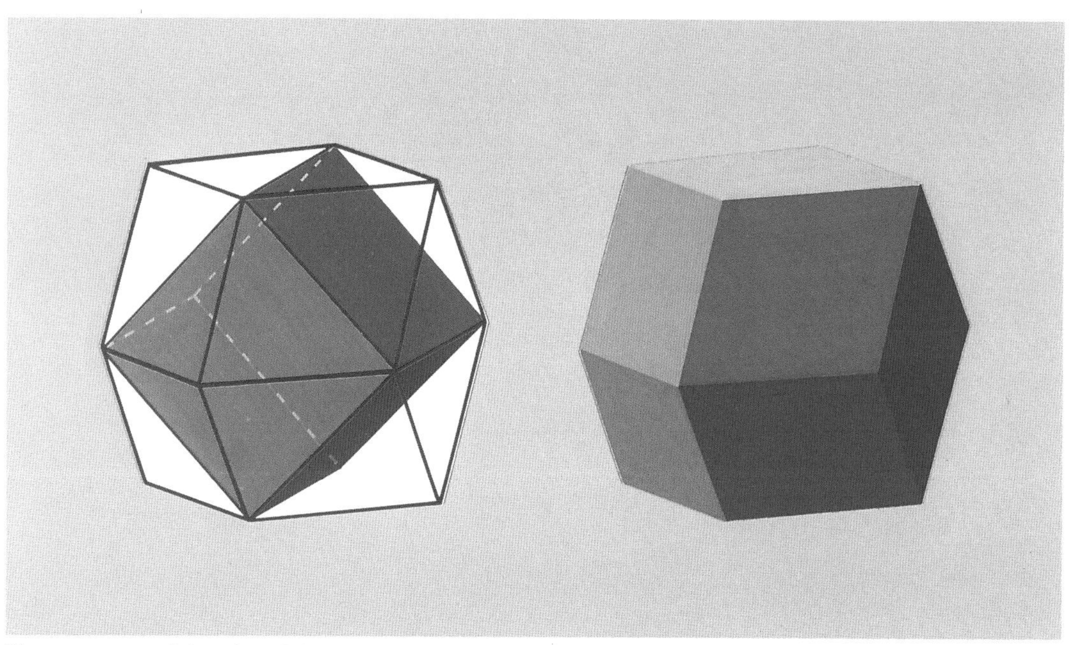

Two aspects of the rhombic dodecahedron.

the centers of the cube faces. This is the figure obtained from the cubic close packed arrangement of spheres if one separates each sphere from its 12 neighbouring spheres by plane faces. Kepler suggested that pomegranate seeds are round to begin with and retain this shape as long as they are small. However, as they continue to grow inside the confined space of the rind, they become squeezed together and adopt the polyhedral shape in order to fill the space as efficiently as possible.

For the layman it is perhaps an amazing fact that the majority of metals crystallize in one or other of these two arrangements, either the hexagonal or the cubic close-packed arrangements of spheres, so that one can actually understand the structure of metals from the Kepler model of highly symmetric assemblies of spherical atoms in mutual contact. But here the reader may harbor the suspicion that this kind of argument is too simple and that just those examples have been chosen where the observed symmetry of crystals happens to fit our somewhat naive assumption of a densest packing of spherical atoms. This suspicion is completely justified. The general case is not quite so simple. The assumption of a spherical shape for polyatomic molecules would be a very crude approximation, about as

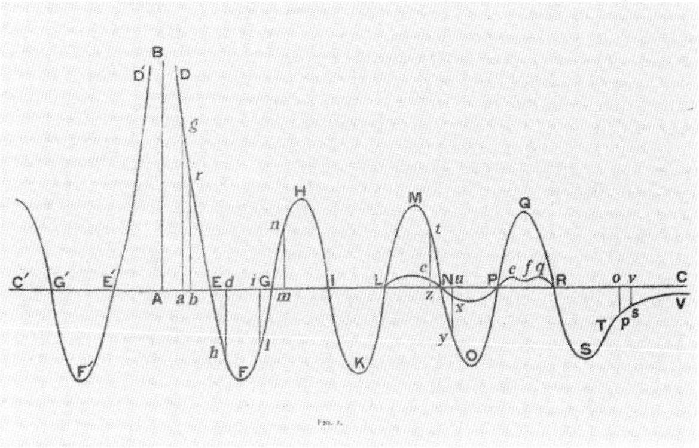

Force between two atoms as a function of distance, according to R. Bošković[9].

unsuitable as the assumption of spherical symmetry for cows as basis for the construction of a cowshed. Apart from such a large-scale idealization of the molecular shape, the spherical molecule hypothesis would also imply that the force with which two molecules attract one another at a fixed distance between their mid-points is independent of whether we rotate the one or the other molecule. It is also implicitly assumed in the Kepler model that this attractive force increases as the distance becomes smaller and is thus greatest when the two spheres are in direct contact, leading to the preference for densest packing. If one wished to explain deviations from densest packing while retaining the assumption of spherical particles, one would be compelled to assume at least that the force between two such spheres depends on the distance in a complicated manner. For certain distances there would have to be attraction, for others repulsion. A suggestion of this kind was indeed put forward by the Jesuit priest Rudjer Bošković (1711–1778)[9]. His idea of the distance dependence of the force between two atoms is shown in the illustration, taken from his *Theoria Philosophiae Naturalis* published in Venice in 1763. With respect to an atom situated at the point A, the force is attractive for distances corresponding to the points F, K, O, S, and repulsive for H, M, Q. (Note that what is here plotted against distance is not the potential energy of the system but its derivative.)

Today we know that this sort of oscillatory behavior is physically unreasonable and we must reconcile ourselves to the fact that the spherical molecule model cannot be salvaged by such ad hoc tricks. Nevertheless, one should not pass over the fact that this attempt of Bošković to rescue the symmetrical spherical model by introducing a specific distance dependence for the force between the particles influenced the thoughts of many 19th century scientists, including, for example, Davy, Faraday, and W. Thomson (Lord Kelvin).

As it happens, molecules in general possess little or no symmetry and their shapes are far from anything remotely recognizable as spherical. Moreover, the forces between them are sensitive to their mutual orientation. Under these circumstances, one may wonder whether 'symmetry', whatever it may mean exactly, can be applied in a meaningful way to arrive at any conclusions whatsoever that are of chemical significance. This will only be possible if one can manage to apply symmetry arguments to problems where at first sight no symmetry appears to be present, where, indeed, the difficulty may lie precisely in the absence of any symmetry. It could also turn out that symmetry is a broader concept than might be inferred from the examples given so far.

IV.

One must classsify things
not from without but from within.

Charles Ferdinand Ramuz

The following elementary examples are intended to illustrate the striking benefits that one can gain from the application of even the very simplest, purely intuitive symmetry considerations. Such considerations can lead to elegant and, above all, surprising solutions for problems which appear at first sight to have nothing to do with symmetry and which can be solved only with much more difficulty by other methods. One might call them, in the current jargon of commercial advertising spots, problems of the type: "Instant insight! All you do is add symmetry".

Example I

Two players, A (who begins) and B (who follows), alternately lay cigars on a table under the sole stipulation that the cigars may not touch one another, as indicated in the following picture for the situation after five moves.

The first player who cannot lay down another cigar under this condition is the loser. Question: which of the two players, A or B, should win?

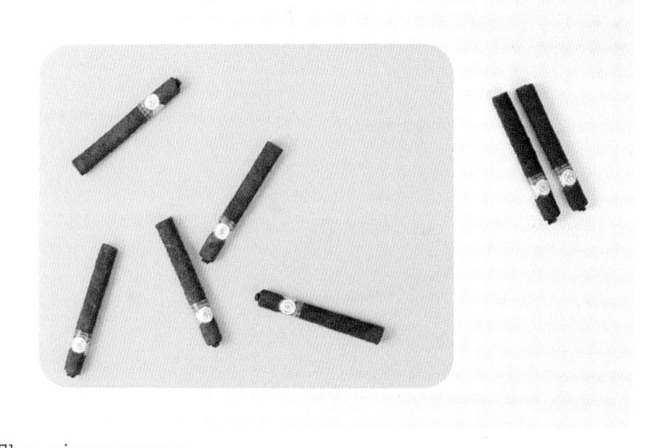

The cigar game.

In view of the complete freedom of position allowed in this game, apart from the rule against touching, an unequivocal answer to this question might seem impossible at

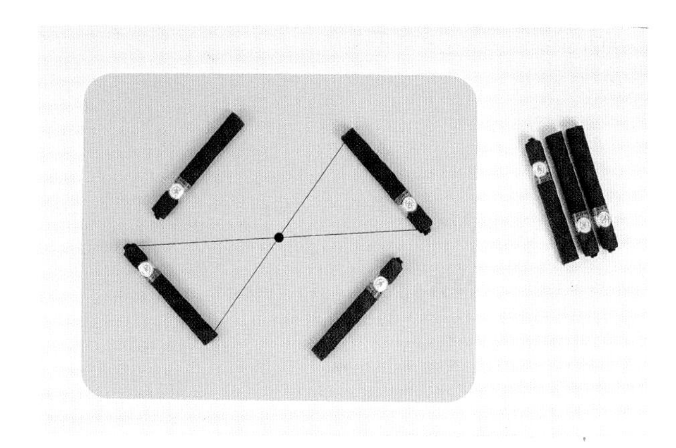

Strategy for the cigar game.

first sight. However, if one symmetrizes the problem, one immediately finds a strategy that appears to lead B to victory. For each cigar laid down by A, B can respond by placing his cigar so that it is related centrosymmetrically to A's with respect to the midpoint of the table, as indicated.

If there is room for A to lay down a cigar then there must also be room for an additional cigar from B. Therefore A is inevitably the first to fail to find room for an additional cigar and must lose.

However, there is one possible first move by A for which there is no corresponding symmetry-equivalent move by B: A simply places the first cigar perpendicular to the table at its midpoint and thereby forces B to place his first cigar at an arbitrary, symmetry-breaking position. From here on, A can always respond to each of B's moves by restoring the symmetry and hence, as the figure shows, stamp him as the loser. By his unorthodox first move, A can thus turn the tables on B.

The symmetrization of the problem and the resulting strategy for A gives the answer to our question: the winner is A, if he remembers to make the right first move.

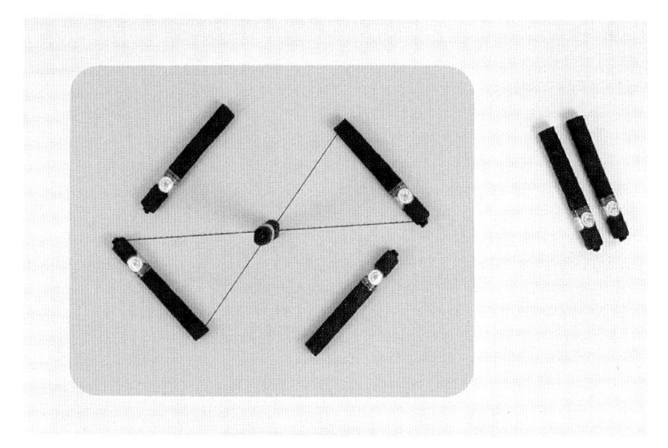

The winning first move !

Example II

The theorem of Pythagoras, that in any right-angled triangle the square on the hypotenuse (c^2) equals the sum of the squares on the other two sides (a^2+b^2), may be branded in the memories of some readers as one of the terrors of their school-days. The usual proof, starting from the pictorial representation of the theorem, depends on the construction of auxiliary lines and on a long sequence of subtle considerations, going back to Euclid, about equal-area rectangles, parallelograms, and triangles, in a manner that many students will remember as rather confusing.

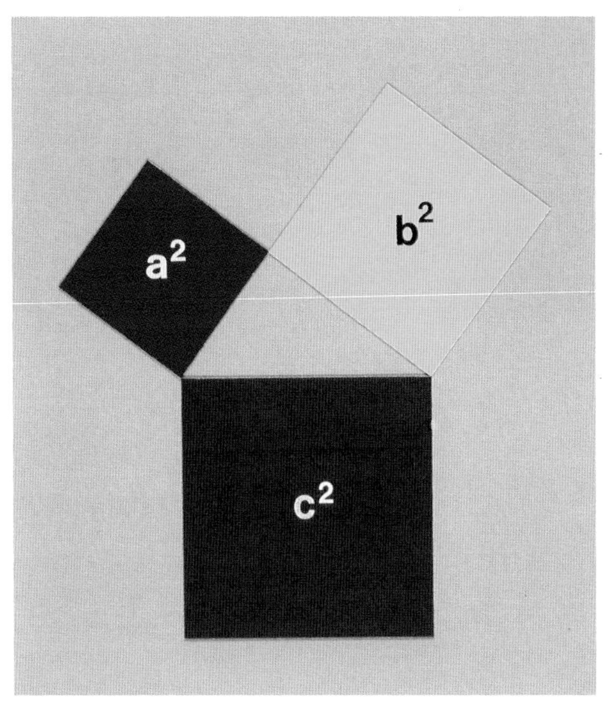

Pythagoras theorem.

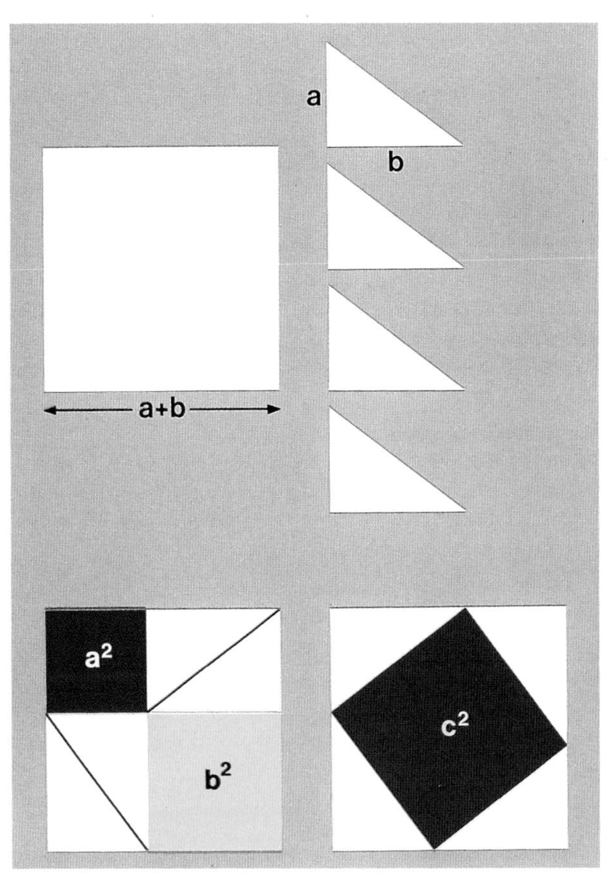

Proof of the Pythagoras theorem according to Bhâskara II.

As shown by the Indian mathematician Bhâskara II (1114–1185?) in the 12th century, one can arrive at a much more lucid proof by symmetrizing the problem. We draw a square with the side $a+b$ as well as four copies of the right-angled triangle with arms a and b. Now we lay the four triangles on the square in each of the two symmetrical arrangements shown in the lower part of the illustration.

Since the area covered by the four triangles must be the same in both cases, it follows that $a^2+b^2=c^2$. The direct insight, obtained by symmetrization of the problem, is so immediate that the Indian scholar provided no explanation to the drawing, except for the laconic instruction: "Look!".

Example III

Consider a scalene triangle and construct an equilateral triangle on each of its sides, to give the geometrical figure shown below. Now show that the midpoints of these three equilateral triangles themselves form an equilateral triangle.

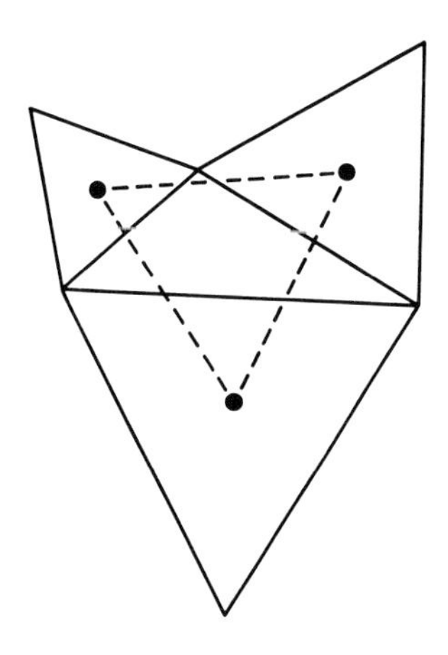

If you try to prove this either by the usual arguments of Euclidian geometry or by using trigonometric formulas, you will probably end up with tedious constructions or lengthy mathematical expressions that are far from simple and not very elegant. On the other hand, the use of a symmetry argument provides a simple and immediately obvious solution.

One only has to recognize that the figure consisting of the four triangles (the scalene one and the three equilateral ones) can be repeated periodically in such a way that it covers the plane. From the Figure it is easily seen that this periodic pattern has threefold rotational symmetry; for example, it is transformed into itself by a 120° rotation about the mid-points of any of the equilateral triangles. It follows that the triangles defined by these midpoints must have also 60° angles, and hence they are equilateral[10].

Tiling of the plane by the above geometrical figure.

Following these three elementary mathematical examples, we now consider three simple, well known problems in physics from a symmetry point of view.

Example IV

The law of the lever says that a system consisting of a thin beam plus two weights, as shown in *A*, is in equilibrium if the product of the weight G_l and the length H_l of the lever arm on the left side is equal to the product of the corresponding quantities G_r and H_r on the right side:

$$G_l \times H_l = G_r \times H_r \qquad (1)$$

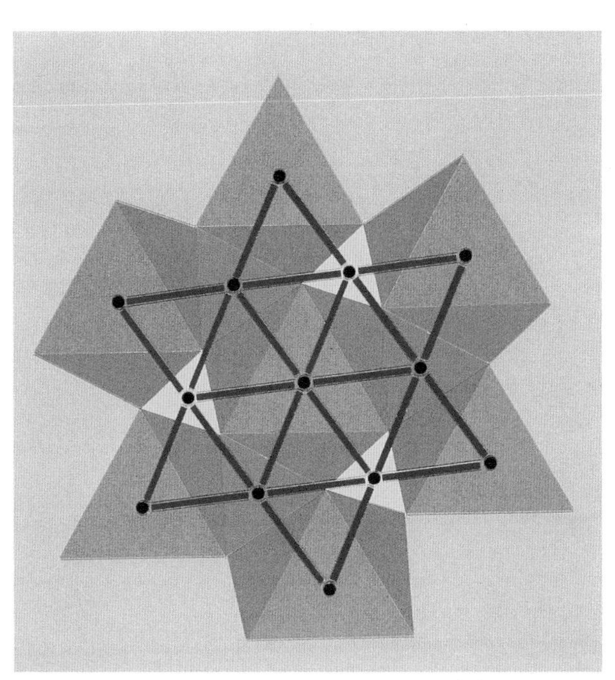

The tiling has threefold symmetry, and the centers of the equilateral triangles themselves make a pattern of equilateral triangles.

The derivation of this law is far from trivial. Even Aristotle was unable to solve the problem, which in his time had already been long recognized as unsolved. It was not until a century later that Archimedes of Syracuse (287–212 B.C.) gave the right answer in his *De aequiponderantibus*.

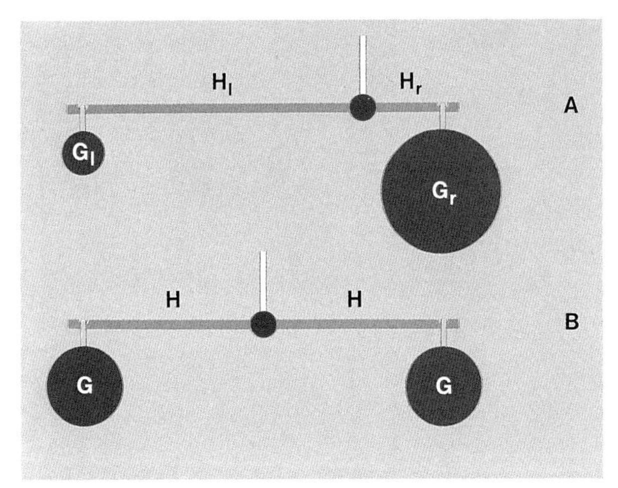

Law of the lever.

There is no doubt that equation (1) must hold for the symmetrical case shown in *B*, since there the system must be in equilibrium 'on symmetry grounds'. Equilibrium clearly holds when both weights are equal ($G_l = G_r = G$) and both lever arms equally long ($H_l = H_r = H$). From this, much later, Galilei Galileo (1564–1642) and independently Simon Stevin (1548–1620) derived a proof of the law (1) in which symmetry considerations play a primary role and which can be followed without difficulty even by those with only a fragmentary relationship to physics and mathematics.

These two authors consider a wooden board suspended at its midpoint and therefore (see *A*), on symmetry grounds, in equilibrium.

The board is now sawn along the dotted lines into three parts: a thin beam *B* and two pieces of unequal lengths $2a$ and $2b$, where $2a + 2b = L$, the initial length of the board. These two pieces and the beam are bored through in such a way that with the help of two threads the original system can be reconstructed, as shown in *B*. It is intuitively obvious that this system must still be in equilibrium if we neglect the weight of the two threads. Since the two pieces are sus-

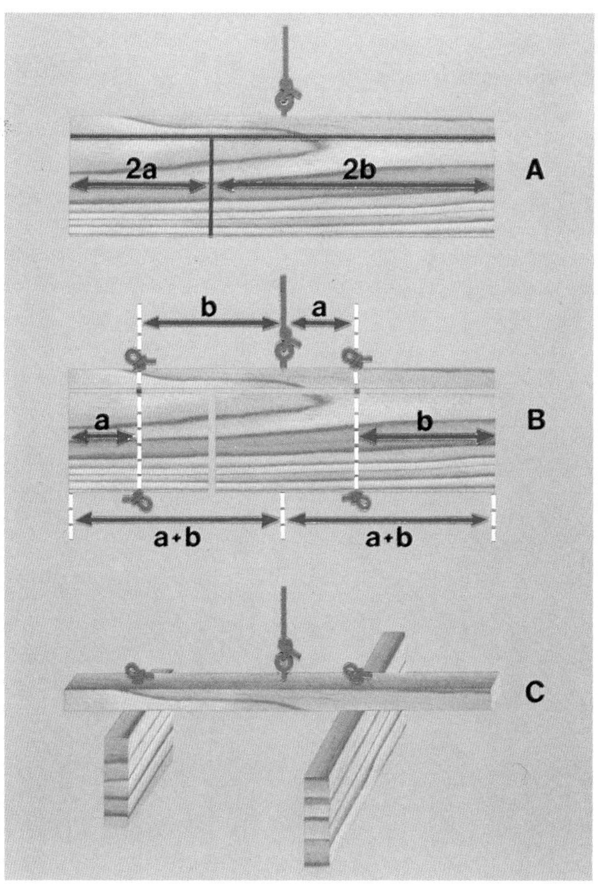

Proof of the law of the lever by a symmetry argument.

pended from the beam only by the two threads, the equilibrium is not disturbed if they are both rotated by 90° to produce the system *C*. Apart from the shapes of the two weights G_l and G_r, this corresponds exactly to system *A*. Now the weight G_l is proportional to the length $2a$ of the left piece, and the weight G_r to the length $2b$ of the right piece, while the lever arms H_l and H_r have the values b and a, respectively, as can be seen directly from *B*. It follows that the product $G_l \times H_l$ is proportional to $2a \times b = 2ab$ and the product $G_r \times H_l$ to $2b \times a = 2ab$. In other words: in equilibrium both products must have the same value, thus establishing the lever law (1).

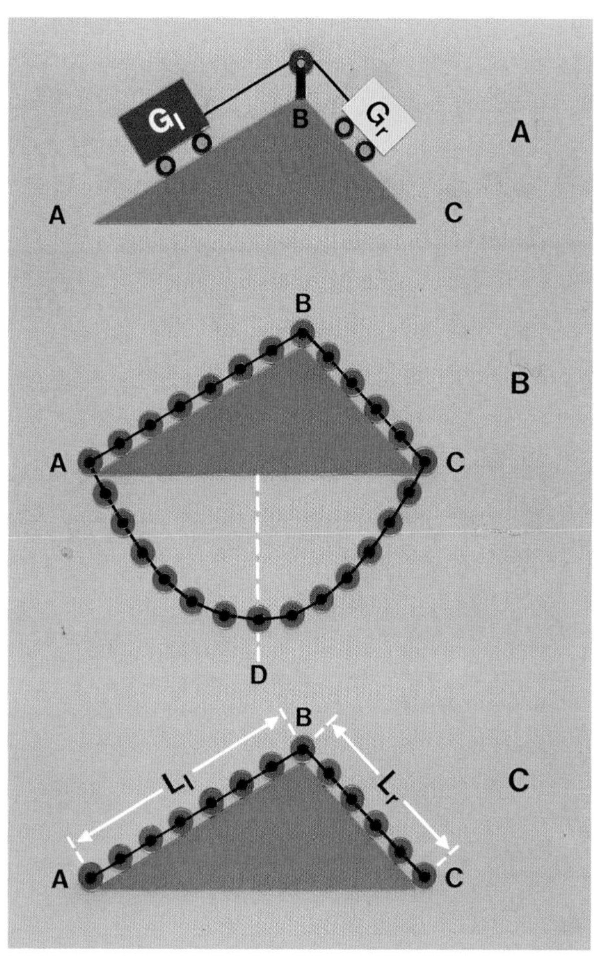

Proof of the law of the inclined plane, according to Stevin.

Example V

The second example (see the Figure) from physics is closely related to the previous one. On the inclined planes AB and BC with equal altitude (see *A*) are two trolleys of weight G_l and G_r joined to one another by rope and pulley. We have to answer the question: when is such a system in equilibrium? An original and astonishing solution to this question was provided by the the Dutch scientist Simon Stevin whom we have already met in the previous example.

In place of the arrangement sketched in *A*, he considered first a pulley-chain of the length ABC, extending over the two inclined planes AB and

BC. The two ends A and C are connected by the additional stretch of chain ADC, hanging downward, as shown in diagram *B*. No one can doubt that such an arrangement is in equilibrium, for if it were not so, the chain would start to rotate in one direction or the other and then continue to rotate forever. In other words: we would have a perpetual motion machine! Since the lower, hanging portion ADC of the chain is symmetrical with respect to a line perpendicular to the midpoint of the edge AC, it must exert the same force at A and C. It can, therefore, be removed without disturbing the equilibrium of the remaining portion of chain lying on the inclined plane. From this, Stevin concluded that the two stretches of chain AB and BC, shown in diagram *C*, must be in balance. But the weights G_l and G_r of these two stretches must be proportional to the lengths AB $= L_l$ and BC $= L_r$, respectively. Thus, Stevin established that equilibrium holds when the condition

$$G_l \times L_r = G_r \times L_l \qquad (2)$$

is fulfilled. If the two stretches of chain AB and BC are now contracted into two masses, represented in the uppermost diagram by the two trolleys of weight G_l and G_r, the equilibrium condition (2) must still hold.

Frontispiece of Stevin's Hypomnemata mathematica.

Stevin was himself so impressed by this proof, which combines the impossibility of perpetual motion with a symmetry argument, that he chose the picture of the closed chain as frontispiece for his book *Hypomnemata mathematica*, published in Leiden in 1605 (Figure, left).

Example VI

Our final example for the simplification of a problem by introducing symmetry considerations can be described, somewhat pretentiously, as an experiment in applied optics. When you look at yourself in a mirror, how large is the outline of your head on the mirror surface? Or, more precisely, what is the ratio of the size of this outline to the actual size of your head? Experience shows that most people have trouble answering these questions, as one can easily convince oneself by posing them (with malice) in a circle of friends. Does the answer not depend on your distance from the mirror? One will even find that the correct answer often meets initially with outright disbelief. You can answer the question experimentally by tracing the outline of your head on the mirror surface.

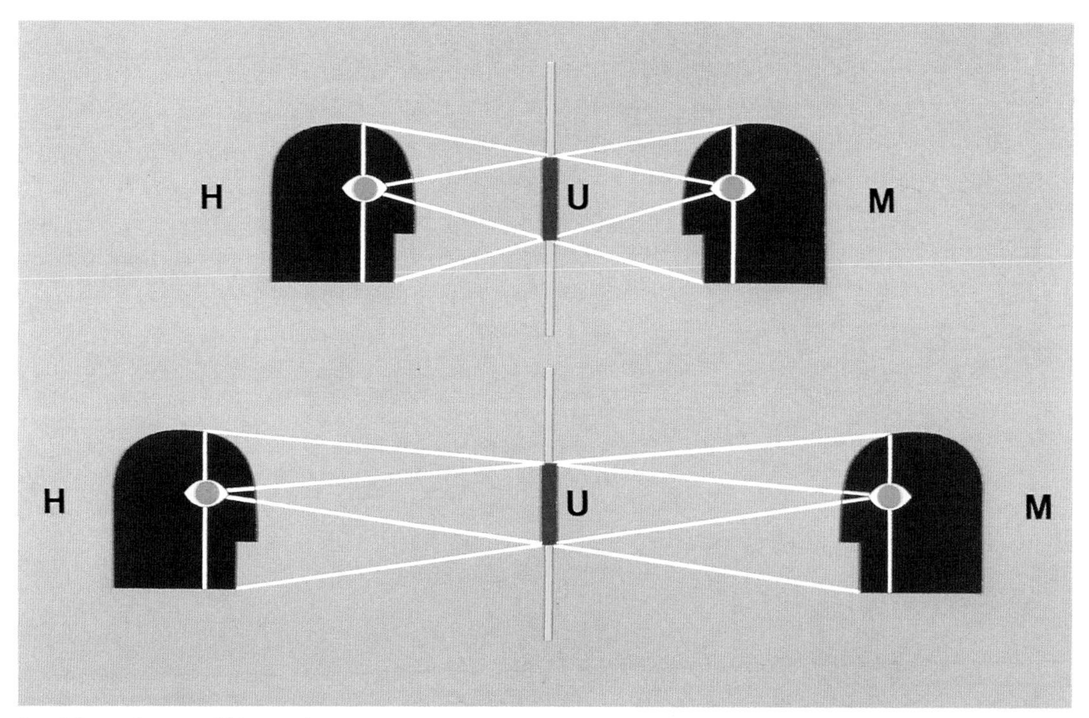

Looking at oneself in a mirror.

Yet a quite elementary symmetry argument leads immediately to the right result. In the preceding Figure, head (H) and mirror-image (M) are symmetric with respect to the plane of the mirror. Indeed, reflection is the very essence of symmetry, at least in our considerations so far. On symmetry grounds, then, M must be the same size as H. As can be seen without difficulty from the diagram, this means that the 'light rays' emerging from the eye of the observer H to the reflected head M must cut the mirror in such a way that the outline U is exactly half as large as H or M.

Moreover, by comparing the upper and lower diagrams, we can see that this result does not depend in the slightest on the distance from the mirror. The correct answer is: the outline drawn on the mirror is exactly half as large as the head itself. Anyone who does not believe it is cordially invited to verify it by suitable experimentation in the privacy of the bathroom.

A provisional stocktaking at this stage might stress the following points:

1. Where 'symmetrization' can be applied to a problem it can lead to a clear and immediately obvious solution.

2. The symmetry argument is often transferable and hence applicable to related problems that look different at first sight.

3. Systematic application of symmetry arguments can lead to a deeper awareness of the relationships among the variables involved in a mathematical or physical problem, even when the symmetry aspect of the problem is not at all obvious.

V.

Rien ne serre autant le coeur comme la symétrie.
C'est que la symétrie, c'est l'ennui,
et l'ennui est le fond même du deuil.

Victor Hugo, Les Misérables, Cosette, 4.1

Where order in variety we see,
and where, though all things differ, all agree.

Alexander Pope

The following elementary examples are intended to illustrate the striking benefits that one can gain from the application of even the very simplest, purely intuitive symmetry considerations.

Before we occupy ourselves any further with the use of symmetry considerations in chemistry, it is necessary to go a little deeper into the underlying concept of symmetry. Until now we have merely assumed that we know by intuition and by feeling what we are talking about, but for more serious, practical applications of the concept that is not enough.

One approach is to define the symmetry of an object, for example, a molecule, by means of 'symmetry operations'. We illustrate this by a simple example. If we rotate a teacup by 180°, as indicated in the adjacent Figure, the initial and final orientations can be immediately distinguished: a cup for left-handers has been converted into a cup for right-handers.

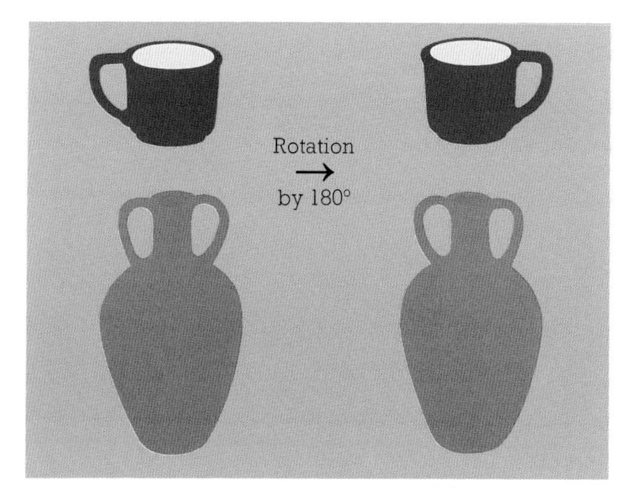

Rotation of a teacup and a vase by 180°.

On the other hand, if we carry out the same operation with a two-handled vase, as indicated in the lower part of the previous picture, the initial and final orientations of the vase cannot be distinguished – assuming, of course, that the vase has no obvious flaws. In other words: if we had left the room, we would be unable to say, on our return, whether the vase had been turned through 180° during our absence or not. Such an operation, which converts an object 'into itself' is called a symmetry operation. In the specific example of our vase, the rotation round 180° is such a symmetry operation. To save words we designate it with the symbol C_2. (The lower index 2 means that only 1/2 of a full rotation, that is, 360°/2, has taken place.) We can say that the vase 'possesses' the symmetry operation C_2, the cup not. More correctly, we should say that the vase possesses a twofold rotation axis as a symmetry element, but for brevity we shall often ignore the distinction between the symmetry element and the symmetry operations that are associated with it.

In addition to the C_2 operation, the vase possesses other symmetry operations. Reflection across a plane that lies either parallel to the plane of the two handles (S) or perpendicular to this plane (S') produces a mirror image that is distinguishable from the original only insofar as it is displaced by an amount that is twice the distance between the vase and the reflection plane, as shown in the upper part of the adjacent diagram. In particular, the mirror image is not rotated with respect to the vase, so that it can be brought into coincidence with the vase by pure translation – whereby we assume in this 'Gedanken-

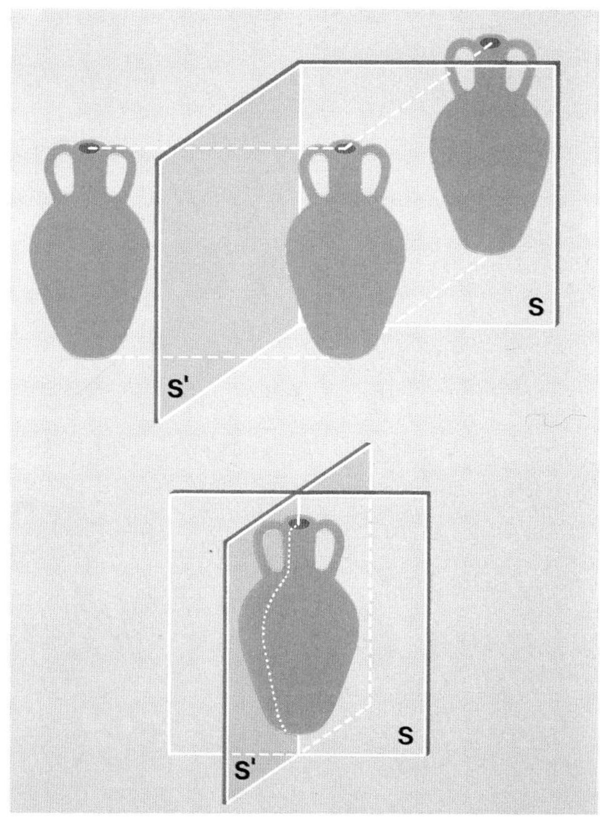

Reflection of a vase.

experiment' that the mirror image can be moved through the reflection plane and out the other side.

In our imagination we can place two mirrors S and S' in the middle of the vase, as shown in the lower part of the previous diagram. In this arrangement, reflection converts the vase into itself. The two reflections thus constitute two additional symmetry operations possessed by the vase, and we describe them in abbreviated form with the symbols σ and σ'.

For technical reasons we always add one further symmetry operation, the identity operation *I*. This operation converts any object into itself, i.e., it does not do anything and may thus appear to be rather trivial. Indeed it is, but its presence serves to guarantee the rule that the result of any sequence of symmetry operations is also a symmetry operation. Perform the C_2 operation twice in succession and you are back where you started; it is the same as if you had done nothing. The operation that, applied to the result of any symmetry operation *M*, produces *I* is called the reciprocal of *M*, designated M^{-1}.

Thus, the vase allows exactly four symmetry operations, namely, *I*, C_2, σ, and σ', while the cup has only two, *I* and the reflection σ. This difference in the number of symmetry operations, four against two, allows us to quantify our intuitive feeling that the vase is 'more symmetrical' than the cup: the greater the number of symmetry operations possessed by an object, the more symmetrical it is.

Another advantage of the method of defining the symmetry of figures by means of symmetry operations is that we can classify together different objects that possess the same symmetry operations as belonging to the same symmetry type. They are then said to belong to the same symmetry group. In this way, the cup, the knife, the dog, and the ethanol molecule shown in the next diagram all belong to the same symmetry group, because they all possess the reflection σ as sole sym-

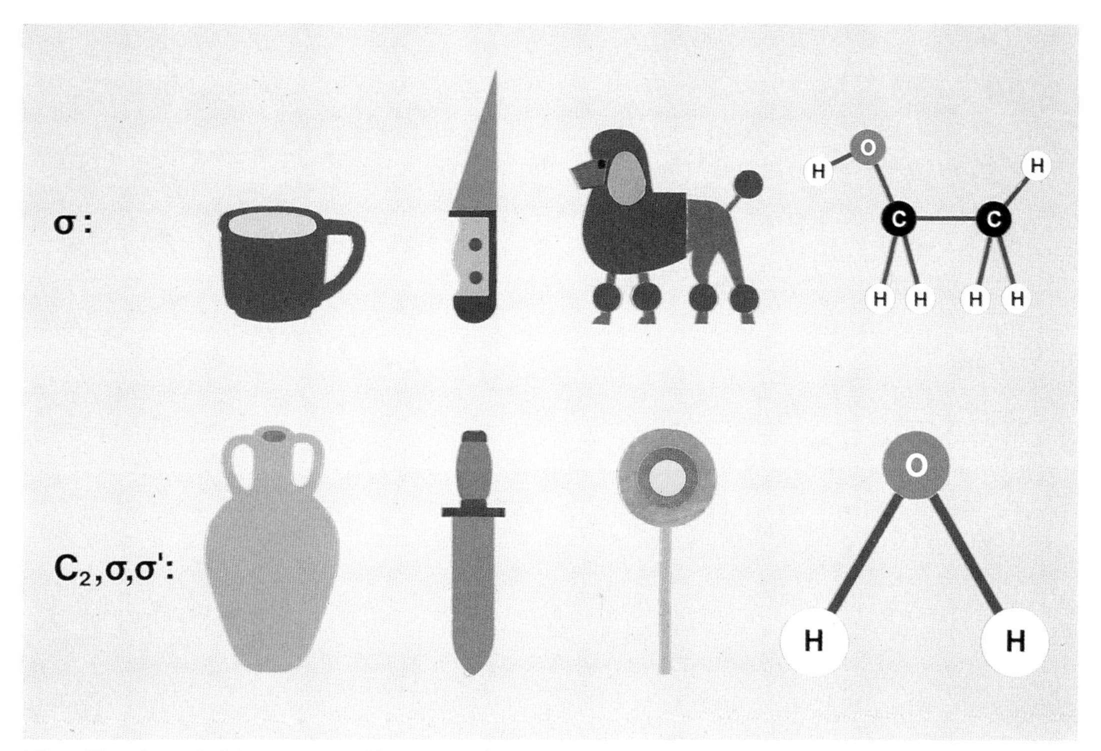

Classification of objects according to their symmetry.

metry operation, apart from the trivial I. On the other hand, the vase, the double-bladed sword, the lollipop, and the water molecule have the symmetry operations I, C_2, σ, and σ' in common. All four belong to the same symmetry group, which is a higher group than that of the first four examples, corresponding to the larger number of symmetry operations that characterize them.

We do not wish to extend our symmetry-operation zoo any more than is absolutely necessary for the understanding of our further exposition. We therefore add only a few to the already mentioned non-trivial symmetry operations, C_2 and σ. The trigonal pyramid, shown in the following Figure, with an equilateral triangle as base, can obviously be brought into self-coincidence by a rotation of 120°, either to the left or to the right. Analogously, we denote this symmetry operation with the symbol C_3, since the amount of rotation, 120°, is equal to 360°/3, one third of a complete revolution. It seems reasonable that we classify C_3 as a higher symmetry operation than C_2, because now a threefold repetition corresponds to a complete revolution. If C_3 is a clockwise rotation of 120°, then repetition of C_3 gives a clockwise rotation of twice 120°, i. e. one of 240°, equivalent to an anti-clockwise rotation of 120°. Thus, whereas the C_2 operation is its own reciprocal, this is not the case for C_3: $(C_3)^3 = I$ and $(C_3)^2 = (C_3)^{-1}$, in an obvious notation. Since the operation has to be repeated three times to produce the identity operation, we say that it is of order three.

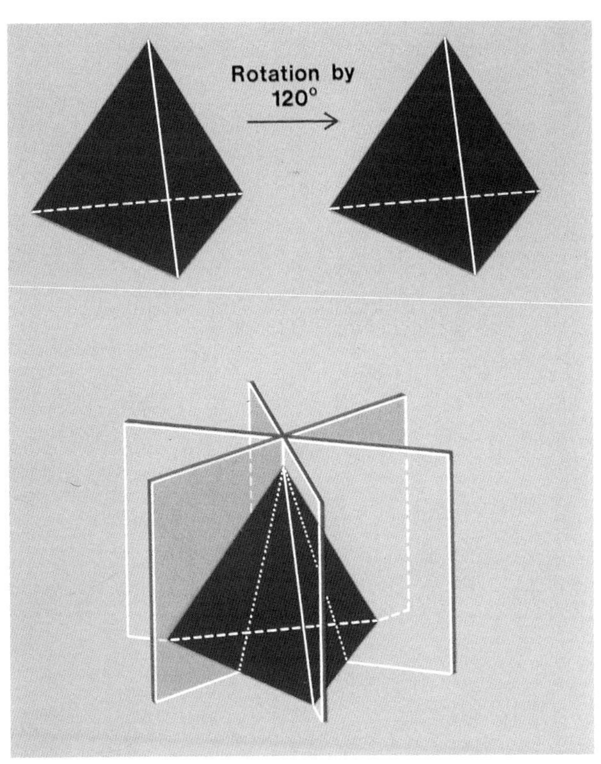

Symmetry operations of a trigonal pyramid.

If, instead of the trigonal pyramid, we had chosen a genuine Egyptian pyramid with a square base as our example, this would be brought into self-coincidence by rotation of 360°/4, corresponding to the symmetry operation C_4. Notice that $(C_4)^2 = C_2$, $(C_4)^3 = (C_4)^{-1}$, $(C_4)^4 = I$. Further along the same track, a five-armed starfish possesses the symmetry operation C_5 and a snow crystal the symmetry operation C_6. It hardly seems worth mentioning that C_6 is a higher symmetry operation than C_5, C_5 higher than C_4, and C_4 higher than C_3.

Returning to the trigonal pyramid, it is apparent that this geometrical figure also shows mirror symmetry. It is brought into self-coincidence by reflection across each of three mirror planes, so that the trigonal pyramid, in addition to I, C_3, and $(C_3)^2$, also possesses σ, σ', and σ'' as symmetry operations; a total of six. On formal grounds, and so that we do not lose track of things, we summarize our symmetry operations in a Table (next page). The symbol is given for each symmetry operation, as well as the order of the operation. If an object or a molecule possesses several symmetry operations, which is often the case, then the order of the overall symmetry is simply the sum of the number of separate symmetry operations, including, of course, the identity operation I.

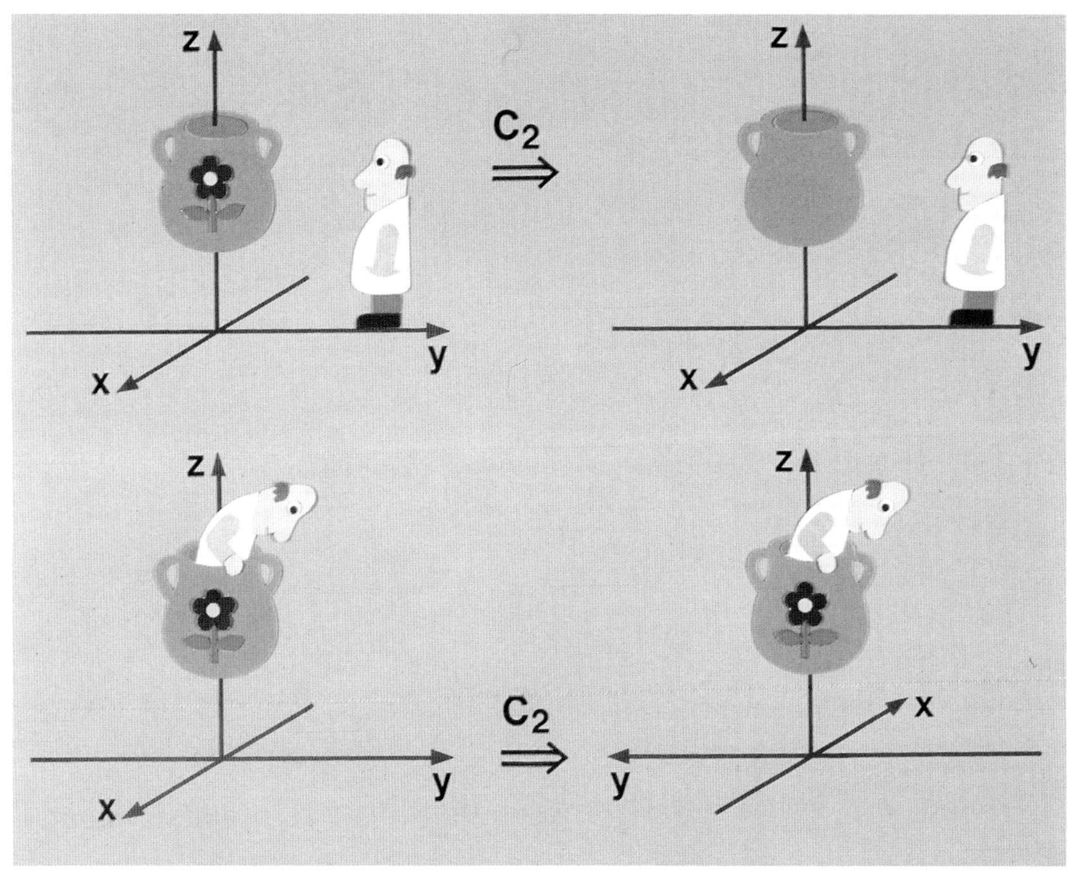

Rotation of a vase by 180° (operation C_2) (top), *considered as a coordinate transformation* (bottom), *depending on the position of the observer.*

Symmetry operations		Symbol	Order
Rotation by	360°/2 = 180°	C_2	2
	360°/3 = 120°	C_3	3
	360°/4 = 90°	C_4	4
	360°/5 = 72°	C_5	5
	360°/6 = 60°	C_6	6
Reflection		σ	2

In what follows, we shall use the elementary concept of symmetry operation as introduced in this chapter. On the other hand, the objects to which these operations refer will from now on be molecules, or more exactly molecular models, which we picture in a tangible fashion as spatial arrangements of atoms linked to one another.

When we talk about rotating an object, our vase say, through 180°, there are two equivalent ways of looking at this operation. As indicated in the previous Figure, we can rotate the object itself or we can rotate the coordinate system with which we describe the object . In three-dimensional space, a reflection is no more than reversing the direction of one of the coordinate axes, a 180° rotation is equivalent to reversing the directions of two of them, and inversion corresponds to reversing all three.

VI.

Irrtümer haben ihren Wert;
jedoch nur hie und da.
Nicht jeder, der nach Indien fährt,
endeckt Amerika.

Erich Kästner

We skip over the alchemical era, when every substance was assigned a separate symbol, and begin around the time when chemists started to think in atomic and molecular terms. This was already the case by 1808, when John Dalton (1766–1844) published his principal work *A New System of Chemical Philosophy*, from which we have borrowed the Table of the elements shown in the Figure below.

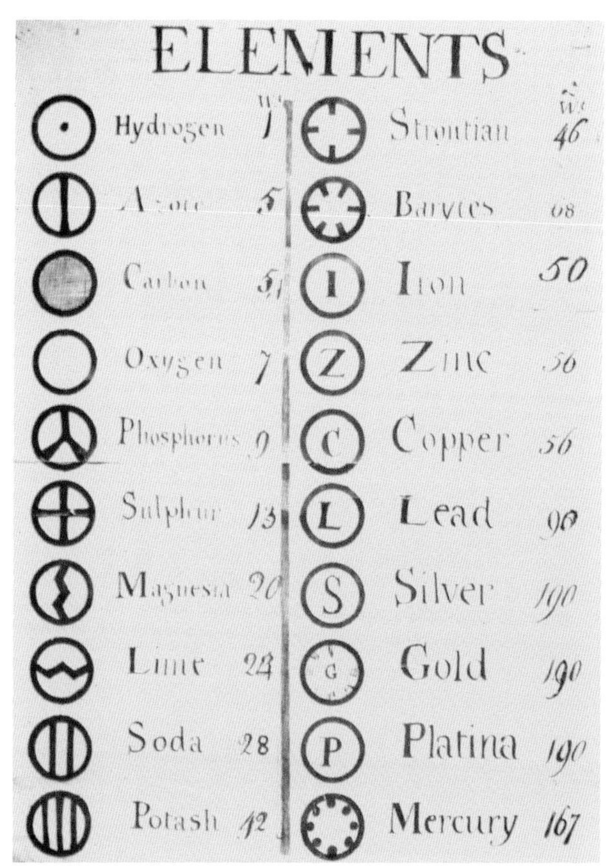

The elements, their symbols and weights according to Dalton[6].

Every circle in the Table corresponds in Dalton's language to a 'simple atom', to which he assigns a weight, shown in the third column, relative to the weight of a hydrogen atom. Out of these 'simple atoms' he then builds 'compound atoms', that is, what we would now denote as molecules. The Figure on the next page, taken from the same work, contains a selection of such Daltonian molecules, including, for example, water (21), ammonia (22), nitric oxide (26), sulfuric acid (31), alcohol (35), and sugar (37). These Daltonian formulas obviously do not correspond to our modern ones. This stems from the fact that the relative atomic weights in the Table are incorrect;

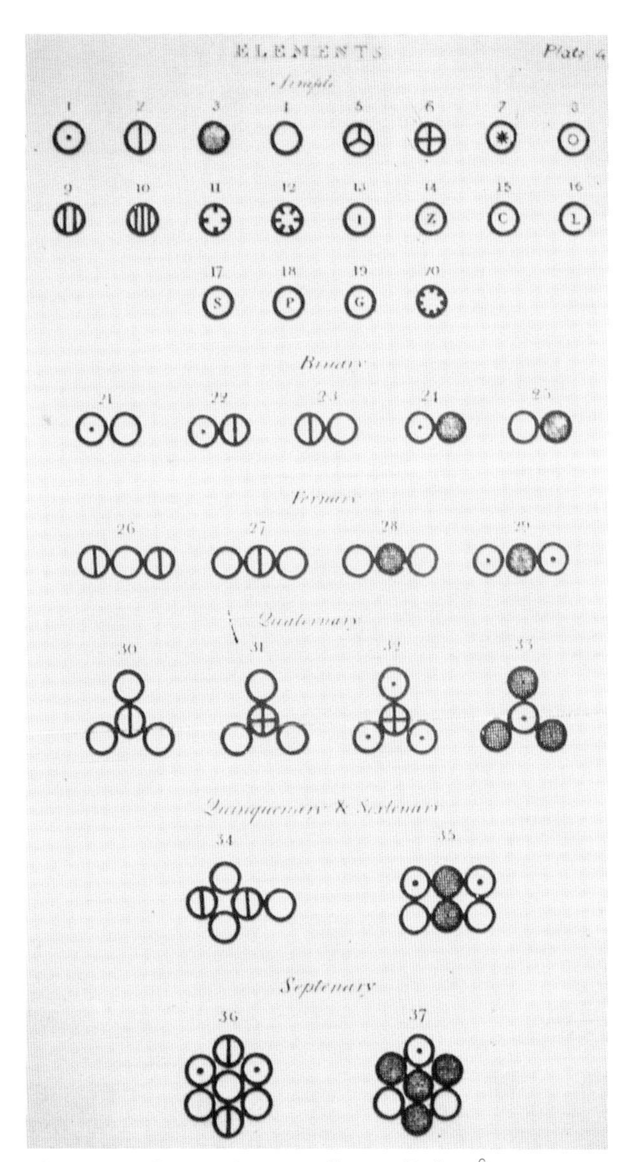

'Compound atoms', according to Dalton[6].

besides that, reliable analyses of most compounds were not available at the time.

Only two of Dalton's basic assumptions are essential for our further considerations. The first is that the 'simple atoms' consist of spherical particles, and the second is that their combination to 'compound atoms' takes place in such a way that the latter are as symmetrical as possible. Here one should not be deceived by the planar representations. Dalton was one of the first who worked with actual molecular models, in which wooden balls, supposed to represent the 'simple atoms', were linked together by rods into the 'compound atoms' or molecules, just as we do today.

In the same year that Dalton's fundamental work appeared, his countryman William Hyde Wollaston (1766–1828) delivered a lecture to the Royal Society[11] in which he postulated that, in making geometric pictures of the relative positions of atoms in molecules, it was absolutely essential to consider all three spatial dimensions. He implicitly assumed that Nature always demands the highest possible symmetry of the molecules formed. For example, if three atoms combine in the proportion 2:1, then the two atoms of the first kind will 'naturally' take up positions on opposite sides of the central atom. If five atoms combine in the proportion 4:1, then, in Wollaston's view, a stable molecule is obtained only when the first four atoms are situated at the corners of a regular tetrahedron with the second type of atom at its center. Presumably, Wollaston was somewhat afraid of his own audacity, for he ended his lecture with the comment that the postulated geometric arrangements of atoms were purely hypothetical. But in his inner convictions he, like Dalton, must have believed in the correctness of his hypothesis, for he added that it was perhaps too much to hope that the geometric arrangement of atoms would ever be known exactly.

This ambiguous attitude, which we have taken into account by putting the words 'atom' and 'molecule' into quotation marks, has its reasons. Although Dalton, Wollaston, and other chemists of that period used chemical formulas – for which the Swedish chemist Jöns Jakob Berzelius (1779–1848) had introduced the familiar letter symbols (H = hydrogen, C = carbon, N = nitrogen, O = oxygen) – quite in the sense of our contemporary models, most chemists were of the opinion that these models had merely a heuristic value. In this connection, the balls of the models or the letter symbols of the formulas were supposed to stand merely as convenient representations for something that was denoted as 'equivalents', without reference to any actual atomic and, especially, spatial structure of matter. By 'equivalents' was meant the corresponding amounts of the elements which could be combined to produce a given substance. We cannot here enter into this controversy, which lasted until the beginning of the present century. Somewhat distorting the history of chemistry, we shall continue as if these early ideas of Dalton and Wollaston had been unopposed in their effectiveness.

In order to discuss molecules in a sensible way, we first need to define some concepts: the sum-formula of a chemical compound, for example,

$$C_3O_3H_6$$

tells us how many atoms of each type are combined in a single molecule of the compound; in this case, three atoms of carbon (C), three of oxygen (O), and six of hydrogen (H). It was once believed that such a sum-formula was sufficient to fully characterize a given compound, and it came as a great surprise when the twenty-year old Justus Liebig (1803–1873) showed in 1823 that silver fulminate (AgCNO), a highly explosive compound, had exactly the same sum-formula as the completely innocuous silver cyanate (AgNCO), which was being analyzed at around the same time by Friedrich Wöhler (1800–1882). In a footnote to Wöhler's paper, Joseph Louis Gay-Lussac (1778–1850), editor of the *Annales de Chimie et de Physique*, made the revolutionary realization that with the same sum-formula the atoms could be linked together in different sequences to give different molecules with quite different properties. He suggested, for example, as one possibility, that the cyanate might be AgCNO while the fulminate might be $Ag_2C_2N_2O_2$, and it is interesting that a continuance of this old but quite incorrect suggestion is still evident in the tabulation of inorganic substances in modern volumes of the *Handbook of Chemistry and Physics*, where silver fulminate is listed as $Ag_2C_2N_2O_2$. The wrong formula has been copied from one book to another for more than 150 years, a time span that contains almost the whole history of chemistry!

Compounds with the same sum-formula were called isomers by Berzelius. The deeper ground for the existence of isomers became apparent only later, when it was realized that, as a rule, each type of atom can only enter into a definite number of 'bonds' with its nearest neighbors: for carbon four, oxygen two, hydrogen only one. This number is called the valency. The quadrivalency of carbon was postulated almost simultaneously around 1858[12] by the Scottish chemist Archibald Scott Couper (1831–1892) and the German chemist Friedrich August Kekulé (1829–1896), whom we shall meet again in connection with the structure of benzene. We can imagine that each atom has the appropriate number of little hooks, which can be linked in pairs to make bonds between the atoms. This picture was developed in France, where even today 'atomes crochus' is a still current expression in a quite different context. In reality, it is somewhat more complicated, but that need not bother us here. Taking account of the valency, we can generally join a given set of atoms into different patterns corresponding to different molecules. Molecules containing the same atoms arranged in different ways are called isomers, a few of which are shown below for the sum-formula $C_3O_3H_6$. They are dimethyl carbonate (**1**), glyceraldehyde (**2**), and lactic acid (**3**).

```
CH3        HC=O        CH3
|          |           |
O          HC—OH      HC—OH
|          |           |
C=O        CH2         C=O
|          |           |
O          OH          OH
|
CH3

1          2           3
```

Notice that multiple (double) bonds as well as single bonds can be formed between pairs of atoms. We simply join two pairs of hooks on each atom. The three isomers depicted represent only a few of the many possibilities for combining the atoms in $C_3O_3H_6$. Formulas of this kind are known as structural formulas, and we emphasize again that they merely show the manner in which the individual atoms are linked together: what we call the connectivity of the atoms in the molecule.

In 1874 the Dutchman Jacobus Henricus van't Hoff (1852–1911) published a little pamphlet[13] which later grew into the well known book entitled *La chimie dans l'espace*. In this book, the idea of the most symmetrical arrangement of atoms in molecules – subliminal in the minds of chemists since Wollaston – was postulated in a bold and quite definite way for the tetravalent carbon atom. Van't Hoff starts with the assumption that the four valencies of carbon are directed towards the four corners of a surrounding tetrahedron (Figure on the next page).

Thus, a highly symmetrical spatial structure is predicted for the simplest hydrocarbon molecule, that of methane, CH_4, as shown in the lower Figure on the next page, translated into a three-dimensional structural formula. To the unbiased reader,

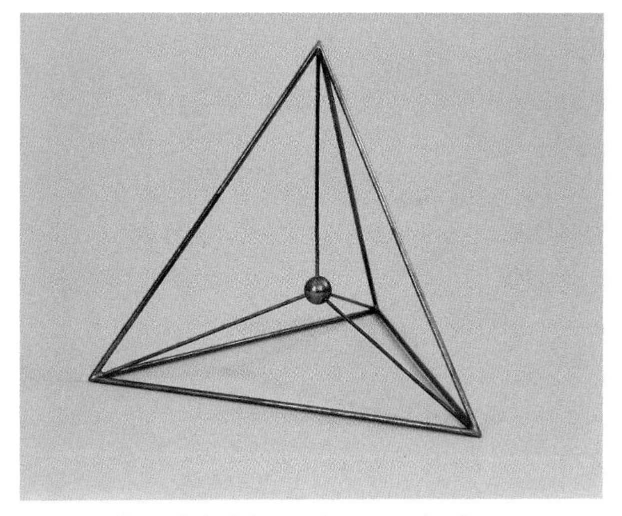

Van't Hoff model of the carbon tetrahedron.

the advance compared with Wollaston's way of looking at things may seem rather modest, but from the resulting spatial models van't Hoff was able to draw far-reaching conclusions that have had a lasting influence on chemical thought and which we shall now discuss in more detail.

Of course, a molecule does not consist of little balls and rods; the individual atomic nuclei are held together by the electronic cloud of the molecule. This cloud occupies the space between and around the individual nuclei, and in order to make a rough picture of this, the chemist today uses space-filling models. A model of this kind is illustrated in the following picture for the methane molecule. The transformation from the planar structural formula to the spatial van't Hoff model and finally to the space-filling model for the lactic acid molecule **3**, which we have already met, is shown in the Figure on the next page.

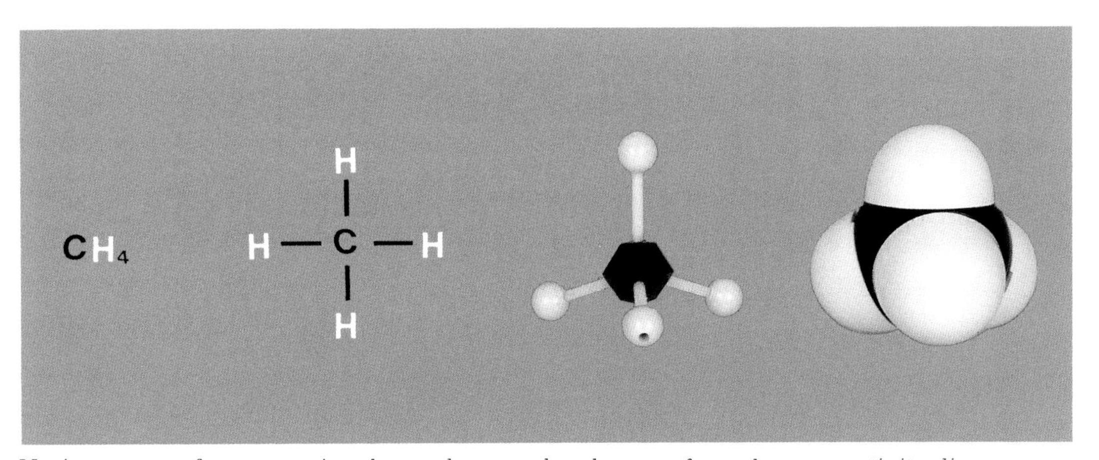

Various ways of representing the methane molecule: sum-formula, connectivity diagram, ball-and-stick model, and space-filling model.

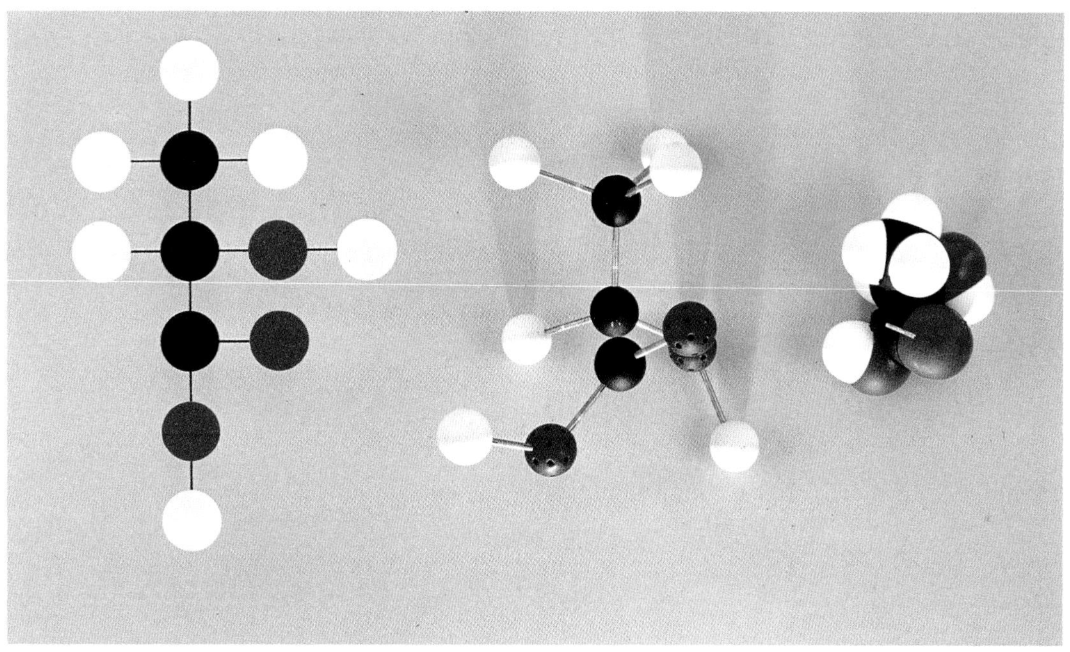

Connectivity diagram, ball-and-stick model, and space-filling model of lactic acid **3**.

Ask a chemist to choose a molecule at random from the round ten million known organic compounds, and it is quite likely to be benzene. Indeed, the importance and historical significance of the benzene molecule can hardly be exaggerated. At any rate, it is especially important from the standpoint of the role of symmetry considerations in chemistry. The problem that arose in the middle of the last century was the following. It was recognized that the benzene molecule consists of six carbon and six hydrogen atoms, hence its sum-formula is C_6H_6. It was also known that substitution of one or several hydrogen atoms by other univalent atoms, that is, those with only a single valency hook, such as chlorine (Cl), led to a quite definite and narrowly restricted number of isomers.

Sum-formula	Number of isomers
C_6H_6	1
C_6H_5Cl	1
$C_6H_4Cl_2$	3
$C_6H_3Cl_3$	3
$C_6H_2Cl_4$	3
C_6HCl_5	1
C_6Cl_6	1

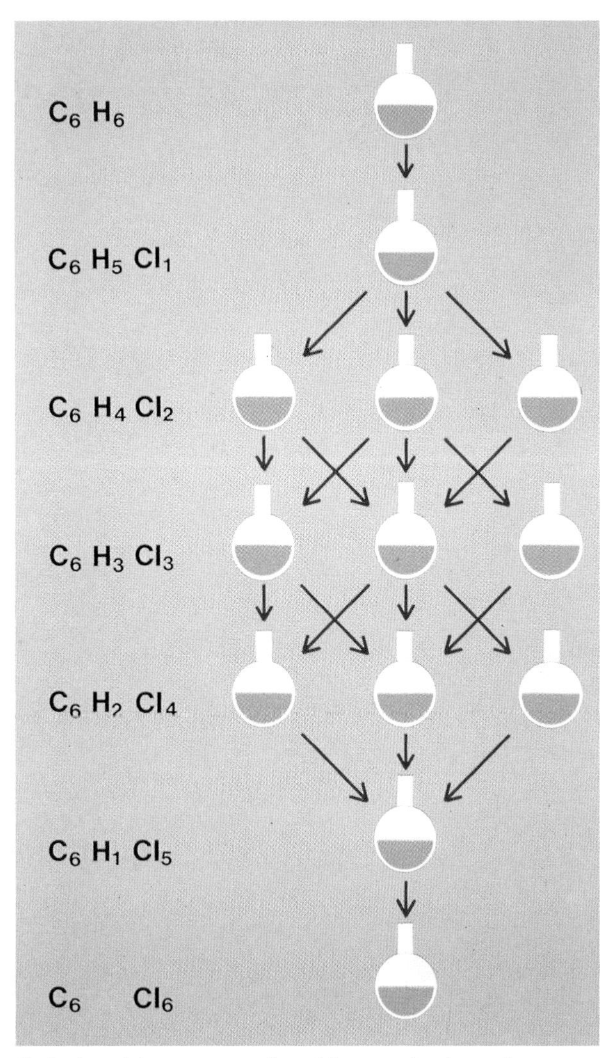

Relationships among the chloro-substituted benzenes $C_6H_{6-n}Cl_n$, $n = 0$–6.

Seen from the chemist's standpoint, this looked in the laboratory something like the picture shown in the adjacent Figure. If benzene (C_6H_6) is allowed to react with chlorine so that only a single hydrogen atom (H) is replaced by a chlorine atom (Cl), one obtains as product only one compound, chlorobenzene (C_6H_5Cl). If this is again reacted with chlorine so that a second hydrogen atom is displaced by a chlorine atom – the chemist says substituted – one obtains this time not just one compound with the sum-formula $C_6H_4Cl_2$ but instead a mixture that can be separated into three different compounds with the same sum-formula but with different properties. There must therefore be three isomers, which require three different structural formulas. Starting with the three $C_6H_4Cl_2$ isomers, substitution of one of the remaining four hydrogen atoms by a further chlorine atom also yields three new $C_6H_3Cl_3$ isomers, as shown in the Figure above. The perplexing thing was, however, that not all of the three trisubstituted isomers could be obtained from each of the disubstituted ones. Thus, one of these isomers gave only a single product, another gave two, and the third gave all three. This connection and the relationship of the higher chloro-substituted benzenes to the lower substituted ones is shown by the arrows in the reaction scheme above. If this scheme is contracted by representing the isomers by points and the reaction arrows by lines, one obtains an abstract figure known to mathematicians as a graph.

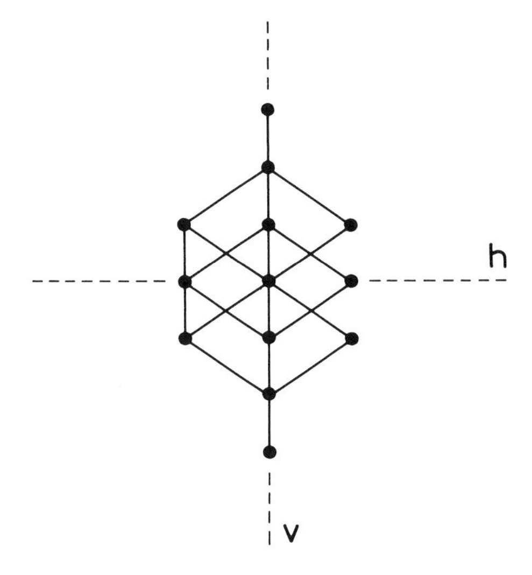

What is new and remarkable here is that we are now not looking at individual molecules, but rather investigating their inherent symmetry through the nature of their interrelationships, symbolized by means of the graph. In the specific example of the chlorinated benzenes, the 'relationship graph' helps us to recognize the symmetry of the reaction pattern especially clearly. The graph is symmetric with respect to the horizontal line (h) but not with respect to the vertical one (v).

Kekulé's own contributions to the benzene problem are reputed to have occurred in circumstances that have become firmly established in chemical mythology[14]. His 1858 recognition of the quadrivalency of the carbon atom is supposed to have happened in a vision, while he was traveling on the upper deck of a London bus from Islington to Clapham Road, and, according to his own later account, he saw the cyclic structure of the benzene molecule in a day-dream while dozing in front of the fireplace in his bachelor study in the Belgian city Ghent in 1861. At least it must have been before his marriage, which took place in 1862. The cyclic formula was actually presented only in 1865, while the story about the daydream was told by Kekulé many years later, in 1890. It is possible that his memory was at fault after such a long interval. According to all accounts, Kekulé was not of a secretive nature; on the contrary, he was in the habit of endlessly discussing his ideas and his latest results with his students and colleagues. It is hard to believe that such an extroverted person could have kept a discovery of that importance a secret for several years, from 1861 until 1865, especially as there seems to have been no obvious reason for reticence. It is more likely that the symmetrical cyclic formula developed in his mind only gradually, starting with the idea of a ring formed by a closed chain of atoms. Indeed, in the 1865 publication *Sur la constitution des substances aromatiques* and also in the lecture that he delivered on January 27th of the same year to the Société Chimique de France with Louis Pasteur in the chair, Kekulé made use of the stretched 'sausage formulas', in which the ring closure is merely symbolized by arrows, as shown in the Figure below.

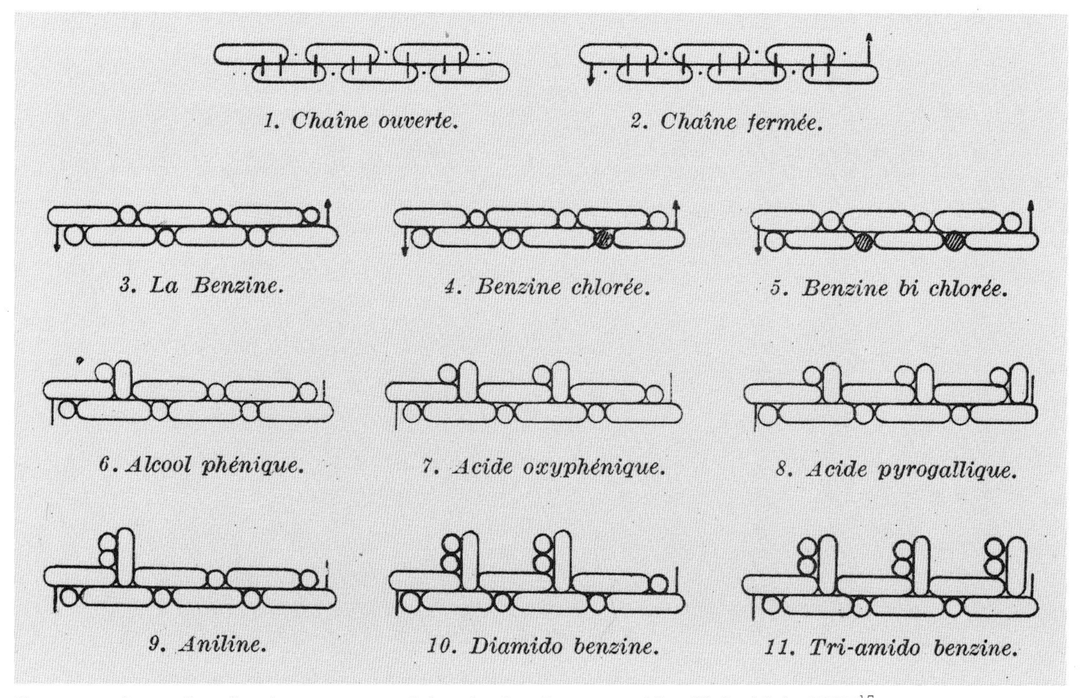

Sausage formulas for benzene and its derivatives used by Kekulé in 1865[15].

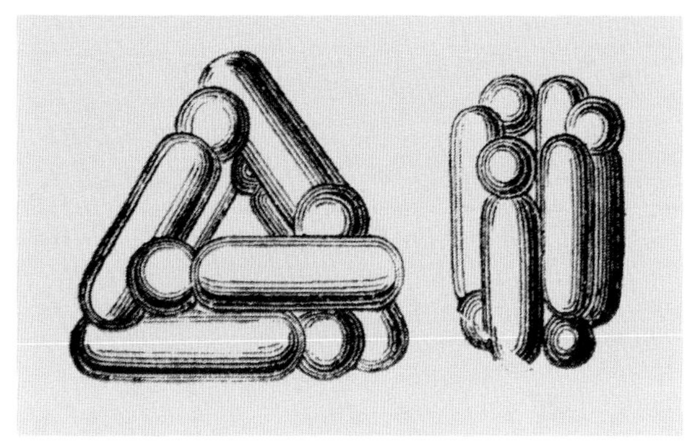

Symmetrical spatial models of the benzene molecule according to Havrez[16].

From these formulas it is not exactly obvious how the symmetry-required equivalence of the six carbon atoms comes about. An attempt to explain this equivalence in terms of a highly symmetric molecular structure was made in 1865 by the now almost forgotten French chemist Paul Havrez, who made use of models with more than a little resemblance to the Kekulé 'sausage formulas' (Figure above). Although these constructions appear from a modern perspective to be beyond good and evil, Havrez's attempt to derive a model in which the atoms in space are symmetry equivalent is quite remarkable.

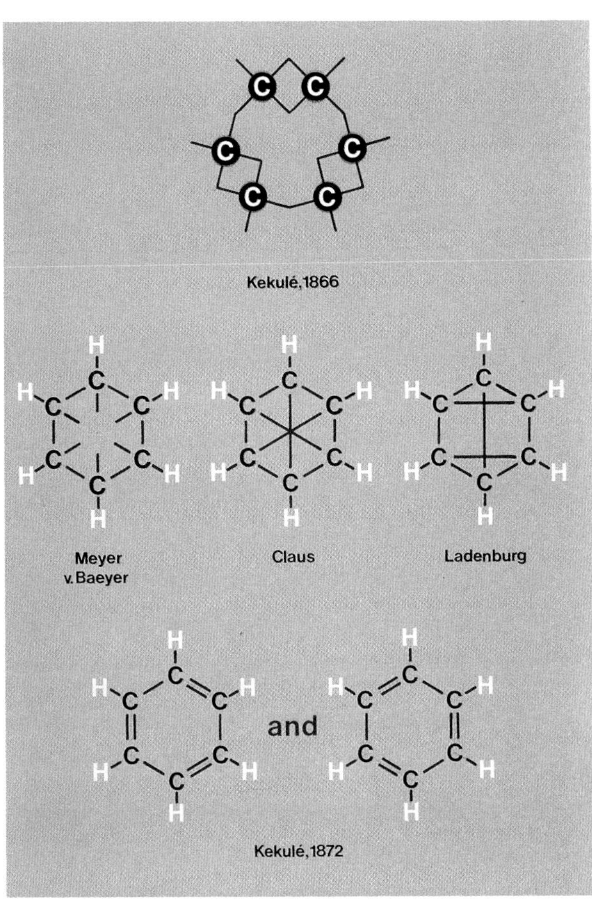

Collection of benzene formulas.

During the period following Kekulé's work, and indeed up to the year 1935, formula after formula was proposed for the benzene molecule, all designed to explain the equivalence of the carbon atoms and also the chemical and physical properties of the compound in terms of the high symmetry of its molecular structure. Some examples are collected in the accompanying Figure.

The first formula links Kekulé's two visions, the one about the quadrivalency of the carbon atom, and the one about the cyclic arrangement of the six carbon atoms in the benzene molecule. To satisfy the rules about joining pairs of valencies to make bonds, one has to assume that single and double carbon-carbon bonds alternate in the ring. This means, however, that the C_6 (rotation of 360°/6 = 60°) symmetry operation is lost, as the formulation allows only C_3 (rotation of 360°/3 = 120°). One unfortunate result of this reduction of symmetry is that it predicts the existence of two isomers in which neighboring hydrogen atoms are replaced, for example, by chlorine atoms. In one of these isomers, **4**, the carbon atoms carrying the chlorine atoms are linked by a double bond, in the other, **5**, by a single bond.

4 **5**

Experimentally, no isomers of this kind have ever been found. As a consequence, other more symmetric formulas were proposed in which this difficulty is avoided by insisting on the presence of the sixfold symmetry operation C_6. The price to be paid was that these formulas do not satisfy the rules of the traditional style of writing; they require the assumption either of 'free valencies' jutting into space or of unrealistically long bonds. Ladenburg's formula is a special case; spatially it

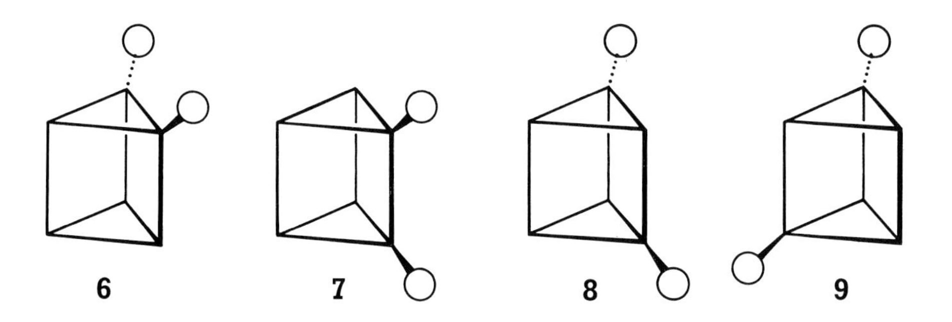

corresponds to a prismatic structure with an equilateral triangle as its base, a figure that gives the correct number of isomers but leads to a different reaction graph from what is actually observed. Ladenburg actually counted four dichloro isomers, **6**–**9**, apparently not recognizing that two of these, **8** and **9**, on symmetry grounds, must be related as mirror images. The reason was that he expected four, as he counted the two isomers corresponding to the Kekulé formulas **4** and **5** as being different. (He also thought that there might be two 1,3 isomers, notwithstanding the fact that rotation by 120° would transform one into the other.)

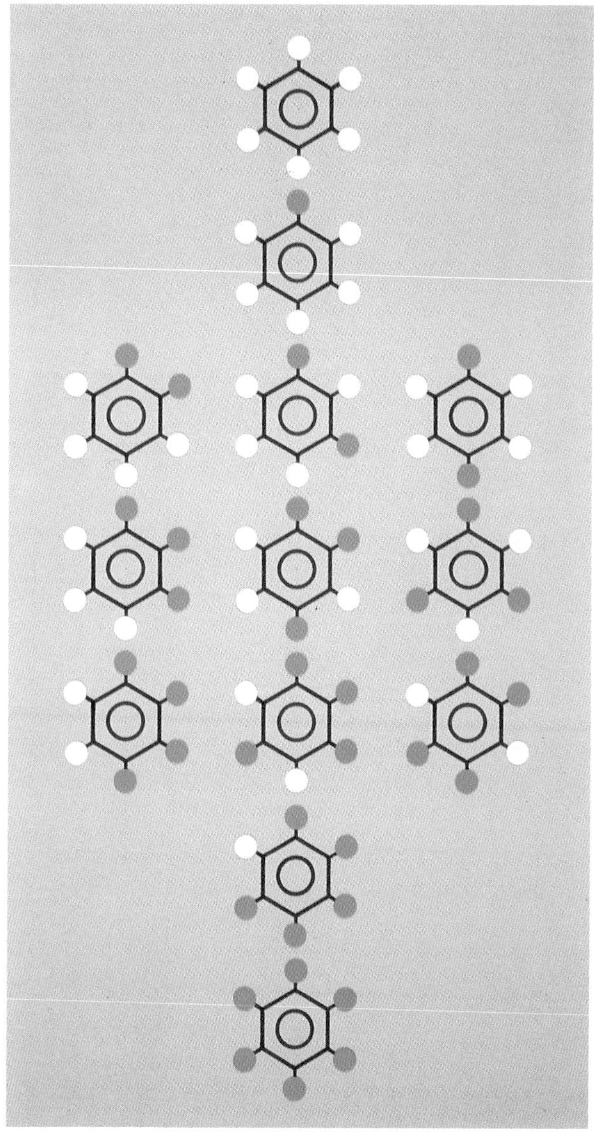

Isomers of chloro-substituted benzenes $C_6H_{6-n}Cl_n$, $n = 0$–6.

Finally, in 1872 Kekulé suggested that his original formula should be kept, but with the important supplementary hypothesis that the single and double bonds continually exchange their positions, as indicated symbolically by the word 'and' in the bottom formula in the Figure on page 49. Thus, the problem of the surplus isomers was eliminated without having to break the rules that were then regarded as valid for formulating chemical compounds. From the Figure above, one can easily convince oneself that successive replacement of white hydrogen atoms (H) by green chlorine atoms (Cl) must necessarily lead to the observed number of isomers as indicated in the Table on page 45, and that, moreover, the graph of the relationships among the isomeric compounds is

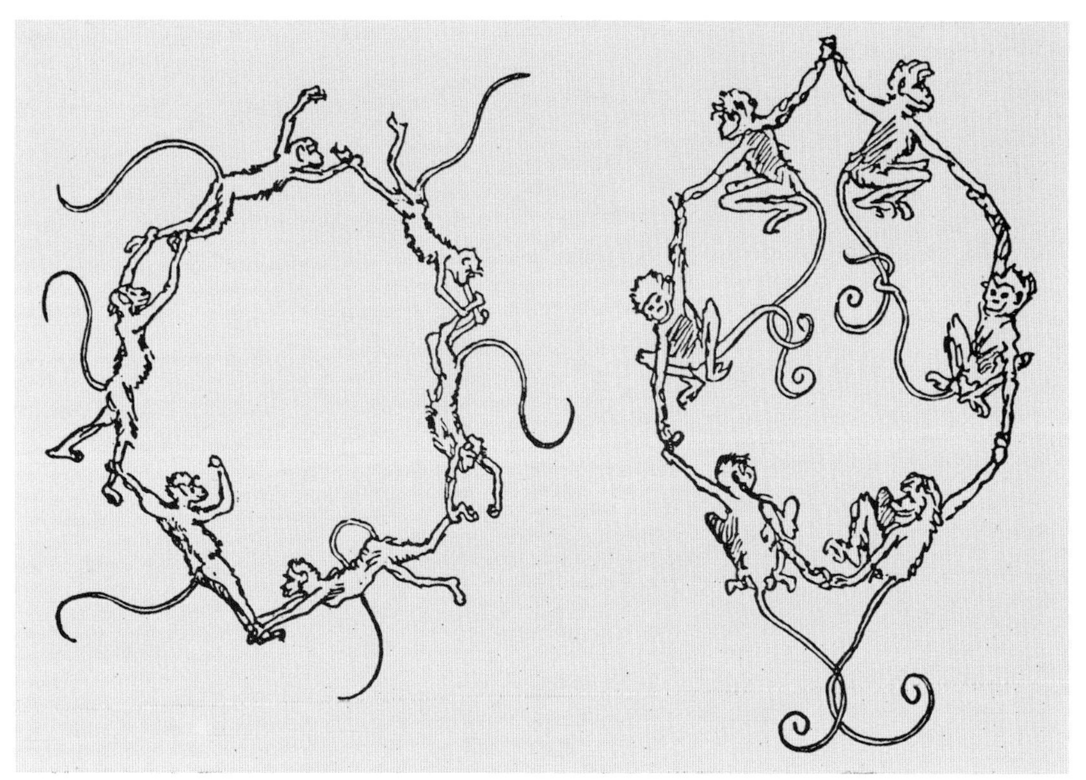

Benzene structure according to F. W. Findig (N.O. Witt).

correctly reproduced. Here one must bear in mind that each structural formula is drawn in an arbitrary orientation, while the corresponding molecule in space can be rotated at will. Thus, for example, in the structural formula of chlorobenzene, C_6H_5Cl, the single chlorine atom can be placed at any of the six symmetry equivalent positions. The molecule in question remains the same.

Kekulé's proposal, forced on him by symmetry, was not regarded by all his contemporaries as the last word in wisdom. This is shown by a parody, presumably written by O. N. Witt under the pseudonym F. W. Findig, in the 1886 humorous special number of the *Berichte der durstigen chemischen Gesellschaft* distributed as a festive gift at a meeting of the German Chemical Society. There it is demonstrated how Kekulé's formula can be derived on a zoological basis; the carbon atoms are symbolized by monkeys whose tails, alternately free and clasped, reproduce the behavior of the double bonds, as illustrated in a drawing of the author (Figure above).

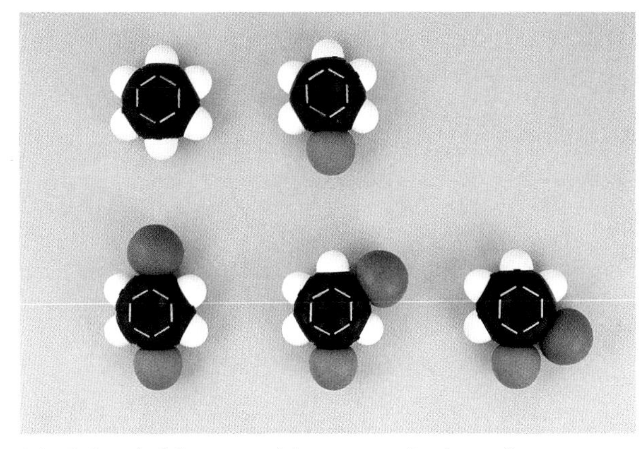

Models of chloro- and bromo-substituted benzenes.

With the chemical and physical methods available at that time, it seemed hopeless to try to obtain any more definite information about the symmetry of the benzene molecule. All the more remarkable is the series of four papers published in the period 1866–1874 by Kekulé's student and collaborator Wilhelm Körner (1839–1925) under the title *Über die Bestimmung des chemischen Ortes bei aromatischen Substanzen*[17]. The problem was to prove the correctness of the symmetrical benzene formula by analyzing the number of isomers and the symmetry of the relationship graph shown on page 47 while maintaining the quadrivalency of the carbon atom. Today it is hardly possible for us to appreciate the extraordinary difficulty of this problem with the then available methods, The point at issue can be explained in terms of molecular models of benzene, chlorobenzene, and the three isomeric bromochlorobenzenes (Figure above). In these, the medium-sized, green balls represent the chlorine atoms, the larger, brown-colored ones the bromine atoms. It is seen that only three isomers of bromochlorobenzene, shown in the lower part of the picture above, can exist. If, in each of these, by some series of chemical reactions, the chlorine atom could be replaced by a bromine atom, and vice versa, then, after the exchange, exactly the same compound should be obtained, provided that the benzene molecule has the hexagonal symmetry assumed by Kekulé. These exchange reactions, and analogous ones with benzene derivatives in which three (or more) hydrogen atoms were replaced by three (or more) other atoms or groups, were intended, taken together, to lead to a proof that all positions in the benzene molecule are indeed completely equivalent. By tackling and solving this problem Körner won a reputation as the philosopher of organic chemistry, but how subtle symmetry considerations in chemistry can be is shown by the fact that the final point in his proof was established only in 1980, more than 100 years later, by the Yale chemist J. Michael McBride[18].

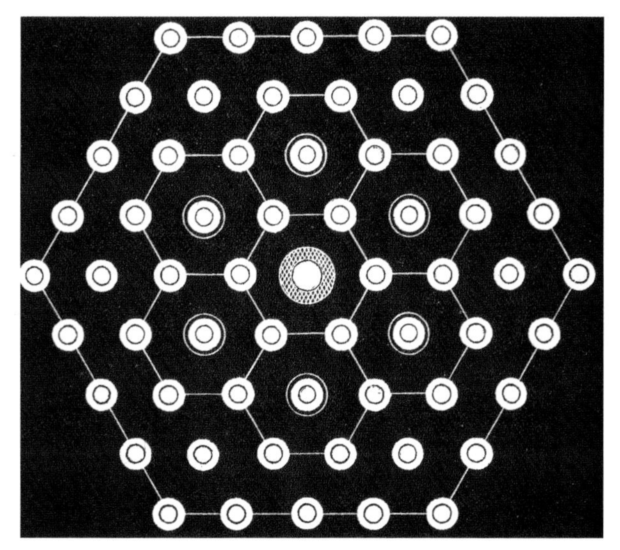

Gaudin's formula for the calcium salt of stearic acid[19].

Scientists too tend to follow fashions. Following Kekulé's proposal of a cyclic structure for a chemical compound, many other formulas showing rings of atoms began to appear. It was unavoidable that some chemists, in their exuberance, postulated cyclic formulas of the highest imaginable symmetry even where these were totally inappropriate. As example, there is the French chemist Marc Antoine Augustus Gaudin (1804–1880), the first to make artificial rubies and sapphires and one of the clearest minds of his era, who in a weak moment proposed the formula shown in the Figure above, a veritable orgy of neo-Kekuléan hexagons, for the calcium salt of stearic acid. In reality stearic acid is a long-chain molecule CH_3–$(CH_2)_{14}$–COOH corresponding to the model shown in the Figure below.

Apart from such extravagances, which obviously originated from the idea that Nature prefers structures of the highest possible symmetry – quite in the sense of the ideas from Plato to Wollaston and strengthened under the impression of

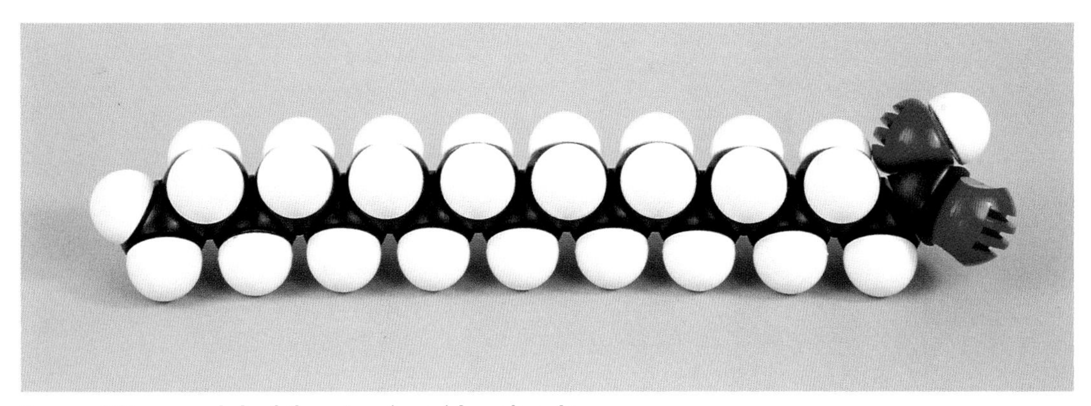

Space-filling model of the stearic acid molecule.

the success of the highly symmetrical benzene formula – this belief in symmetry as a principle could also lead to false predictions, even in cases where its validity might have appeared fairly secure.

The attempt to generalize Kekulé's benzene formula led inevitably to the search for the eight-membered analog, cyclooctatetraene **10**. This molecule ought to consist of eight carbon atoms arranged in a ring with alternating single and double bonds. Extrapolating from the chemistry of benzene, and with the implicit assumption that here too the principle of maximum possible symmetry must apply, it was expected that the two 'Kekulé formulas' below ought to provide a correct description of the molecule.

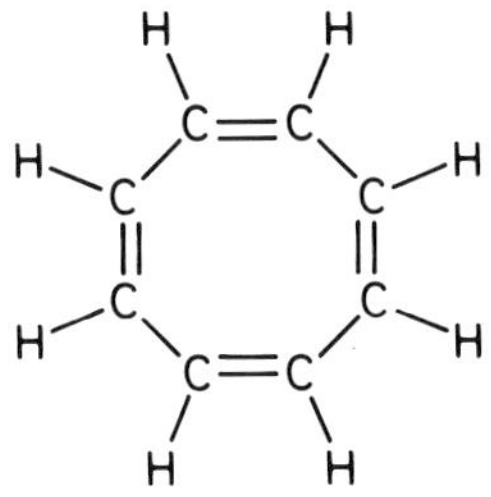

and

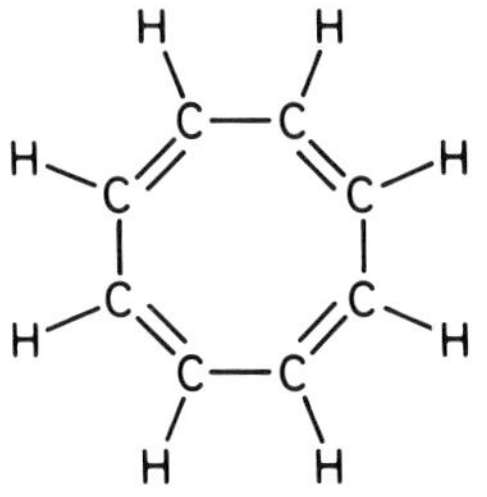

10

From this formula, cyclooctatetraene was expected to behave as a regular octagonal planar molecule, with chemical and physical properties very similar to those of benzene. In 1911 Richard Willstätter (1872–1942) succeeded in obtaining the compound, and it was a great surprise when it turned out to have quite different properties from those expected – it is very unstable and has not the slightest in common with benzene. In particular, as was found later, the molecule does not have the highly symmetric structure shown on the left of the next Figure[20], which had been taken more or less for granted, but a non-planar structure of lower symmetry, as indicated by the model on the right of the next Figure. The deeper ground why this molecule has a lower symmetry than its maximum possible symmetry is quite subtle and was recognized only in the middle of the 1930's by the German physi-

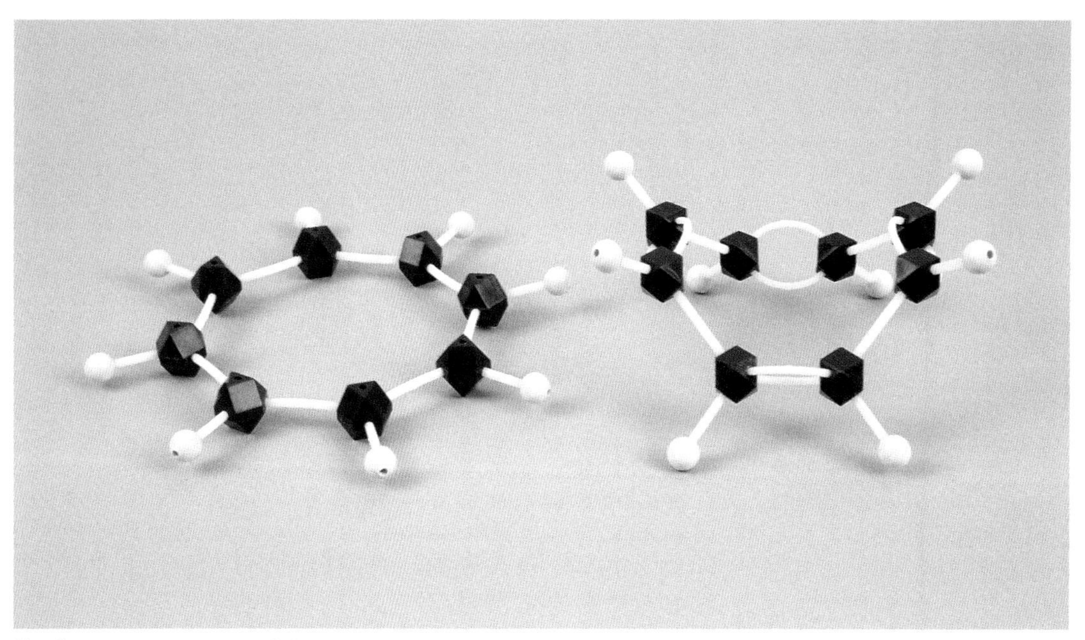

Cyclooctatetraene models.

cist Erich Hückel[21] (1896–1980), one of the fathers of modern quantum chemistry.

Another example of how a psychological 'symmetry fixation' can lead to false conclusions is provided by the cyclic saturated hydrocarbons. They are called 'saturated', because they contain no multiple bonds (unlike our benzene formula), but only single bonds between carbon atoms (C–C) or between carbon and hydrogen atoms (C–H), so that the four valencies of each carbon atom are used to make bonds to four separate atoms. The first seven cycloalkanes with three to nine carbon atoms in the ring are shown in the following Figure.

It was soon discovered that in the organic compounds found in Nature, such saturated rings occur with quite different frequencies. While the five- and six-membered rings are very widely distributed, the seven-membered ring is found only rarely and the others, the three-, four-, eight-, nine-, and higher-membered rings, practically never. In 1885, Adolf von Baeyer[22] (1835–1917) gave what looked like an appealingly simple explanation for this observation. With the assumption that all cycloalkanes have the highly symmetric planar structures shown in the following diagram, it is easy to

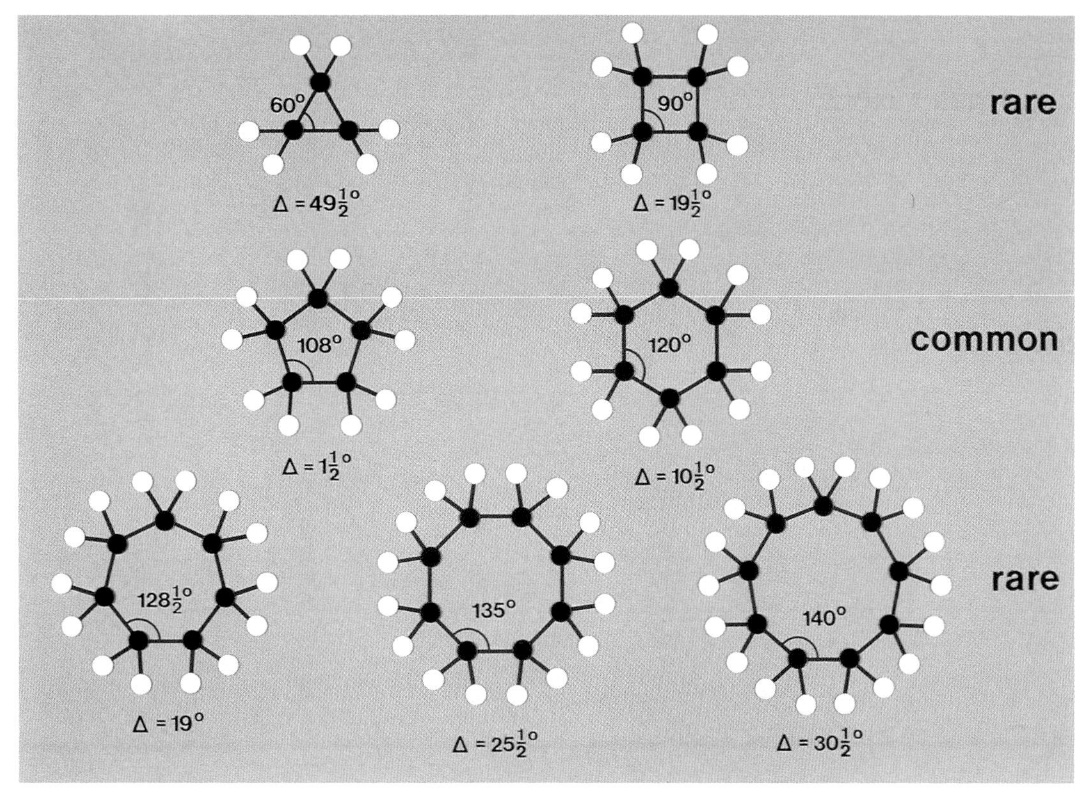

Bond angles in planar cycloalkane models.

calculate the angle between adjacent C–C bonds. Now the 'natural' directions of the four valencies of a carbon atom, for example in the methane molecule, are such that the angle between two bonds emanating from this atom make the ideal angle of 109.5° (the angle between lines drawn from the centre of a regular tetrahedron to its corners). In order to obtain the angles that occur in these hypothetically planar cycloalkanes the bonds must be bent either inwards or outwards by quite definite amounts Δ. These Δ magnitudes are shown underneath the formulas in the Figure above. If the individual bonds were to bend like springs, it is clear that, except for the five- and six-membered rings, a considerable strain would arise. The further the deviation from the ideal angle of 109.5°, the more the angles would have to be deformed, and the greater the strain would be. This quantity, named Baeyer strain after its inventor, seemed to be a plausible and sufficient explanation for the observed frequency of occurrence of the various ring types in Nature. The only drawback was that it later proved to be quite incorrect for rings containing more than three carbon atoms, because these do not occur as planar rings, as had been tacitly assumed 'on symmetry grounds'. The only cycloalkane for which the carbon ring turns out to be planar is the one that cannot avoid it – cyclopropane, with a three-membered ring. The others are all puckered, more or less.

In 1890 Hermann Sachse[22] (1862–1893) had pointed out that two puckered, 'strain-free' forms of cyclohexane are possible: a 'symmetrical', **11**, and an 'unsymmetrical', **12**, form, corresponding to what we now call the chair and boat forms. As Sachse observed, there are two kinds of positions for substituents in the chair form, which are interconvertible by a ring inversion process. He also noticed that the boat form is flexible, whereas the chair form is rigid, provided that the valency angles remain constant.

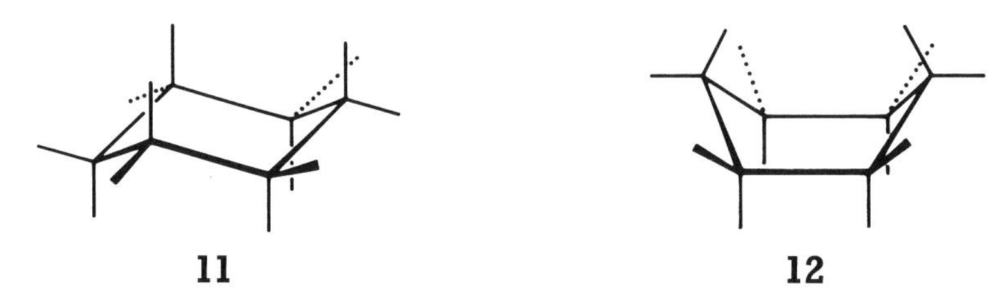

Sachse's views about the cyclohexane structure did not meet with acceptance, partly because they were at variance with Baeyer's strain theory, and partly for a deeper reason. They predicted the existence of at least two isomers of a mono-substituted cyclohexane derivative, while many efforts to obtain more than one such isomer were uniformly unsuccessful. However, in 1918 Ernest Mohr[22] (1873–1926) revived Sachse's long neglected model; he pointed out that the valency angle of 120° required by a planar hexagonal ring was unlikely to be adopted in a cycloalkane molecule, and that rapid inversion of the chair form might explain the experimental failure to isolate more than a single isomer of a mono-substituted derivative. The isolation of two forms (*trans*, **13**, and *cis*, **14**) of decahydronaphthalene (or decalin for short) was a severe blow to the adherents of the Baeyer theory, and, by the mid-1920's the Sachse theory of puckered, strain-free rings was more or less accepted although it was still not known whether cyclohexane existed in the symmetrical chair form or the flexible form, or as a mixture of both.

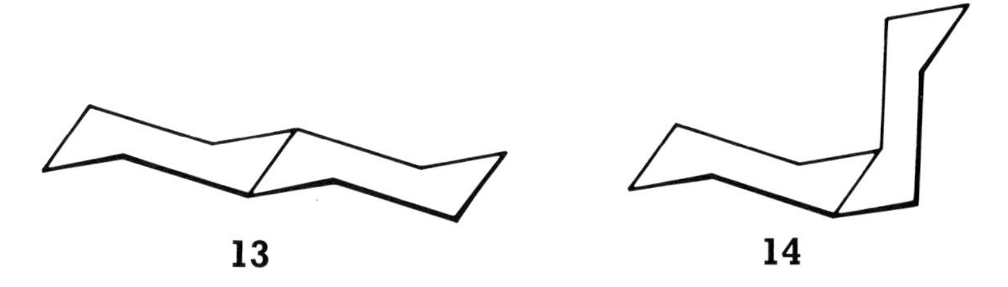

There is an amusing epilogue to this story. With the advent of crystal structure analysis (to be discussed later in this chapter) this method was applied in 1926 to crystals of hexachloro- and hexabromocyclohexane[23]. From the results, it was quite clear that both molecules must have highly symmetrical structures; a threefold rotation axis combined with a centre of inversion, which would be compatible with the planar form and with Sachse's symmetrical chair form but not with his unsymmetrical form. Since, by this time, the planar form could be excluded, this left the chair form as the only possibility. It is therefore quite puzzling to read the authors' own account of their research: "It is of interest to note that Mohr's theory of strainless rings.... is not compatible with our conclusions .....His three-dimensional formulas have a center of symmetry in one case, but not a plane of symmetry. Four carbon atoms of a given cyclohexane ring are coplanar, the 1, 4 carbon atoms being equidistantly placed above and below the plane. Such a representation would not be tenable on the basis of our conclusions".

One can only guess that the authors had not looked at either Mohr's or Sachse's original papers (mentioned in footnotes) but had read the background material in a review article published in the 1924 *Annual Reports of the Chemical Society* (also mentioned in a footnote). There the English chemist Christopher Ingold (1893–1970) reviewed the growing evidence for the Sachse-Mohr theory and illustrated his arguments with diagrams similar to **15**, of the kind that could be set by ordinary printer's type. Is it possible that these diagrams were interpreted as indicating puckered rings with unequal interatomic distances and lacking a threefold rotation axis? We shall never know. In a second paper on the crystal structure analysis, no mention was made of the somewhat paradoxical statements of the first paper; this time the evidence was interpreted as establishing the chair form **16** of the ring, with the substituents in the equatorial positions.

The dogma of the instability of the many-membered rings was finally broken by Leopold Ruzicka (1887–1976). In connection with studies of the olfactory components of musk, he synthesized a whole series of compounds, in which up to 30 carbon atoms were linked in a ring. These rings are neither highly symmetrical nor planar, as Baeyer had assumed; rather, they are built from two parallel zigzag chains, linked to each another at both ends, as seen in the space-filling model of the many-membered cyclic ketone shown in the next Figure.

Benzene rings can be fused together in all sorts of ways, and chemical compounds corresponding to many of these combinations have been made. In fact, some of them are rather familiar, such as naphthalene (**17**) and anthracene (**18**). The structure with a benzene ring fused to six other benzene rings has a high symmetry, the same as that of benzene itself, and is called coronene (**19**). If one continues the process of fusing benzene rings together

Space-filling model of a hypothetical structure of a many-membered cyclic ketone.

indefinitely, one arrives at an infinite hexagonal sheet of atoms (**20**). Graphite, the stable form of carbon, consists of stacks of such hexagonal sheets. The ease with which these sheets can slide over one another is responsible for the lubricating properties of graphite.

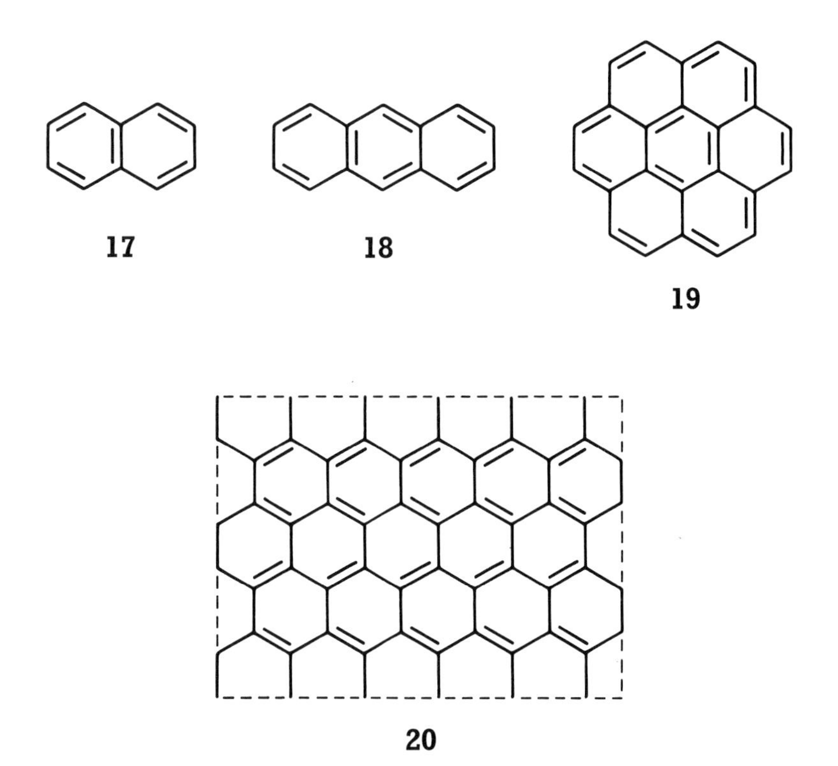

Diamond, the other well known and more expensive form of carbon, also has a highly symmetric repeating structure in which each atom is surrounded by four others at the vertices of a regular tetrahedron (Figure on the next page), in perfect agreement with the model proposed by van't Hoff. A crystal of diamond can be regarded as a giant three-dimensional molecule, in which all the atoms are held together by chemical bonds. It is this property that is responsible for the exceptional hardness of diamond – it is actu-

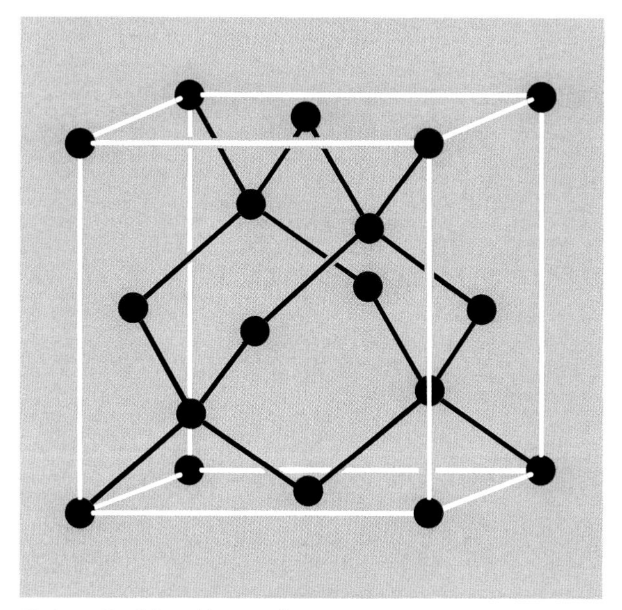
Unit cell of the diamond structure.

ally the hardest known substance. Of course, diamonds have been known as highly prized gemstones since antiquity. But the claim made in advertisements that 'diamonds are forever' is exaggerated. A diamond is actually less stable than the more humble form of carbon. Wait a few million years and a diamond is quite likely to transform into graphite.

A more recent example where an unqualified belief in maximal symmetry at all costs would have led to the correct answer is provided by the 60-atom cluster of carbon atoms, **21**, discovered in 1985 by Richard E. Smalley and Harold Kroto, and the object of much attention since then[24]. This cluster can be made, along with smaller amounts of other clusters containing different numbers of atoms, by vaporizing graphite in a helium atmosphere either with a laser beam or by electric heating. But the 60-atom cluster predominates. Why 60? As a bright student of symmetry might guess, and a soccer enthusiast would probably know, the answer is that 60 equivalent points can be placed on the surface of a sphere, and in such a way

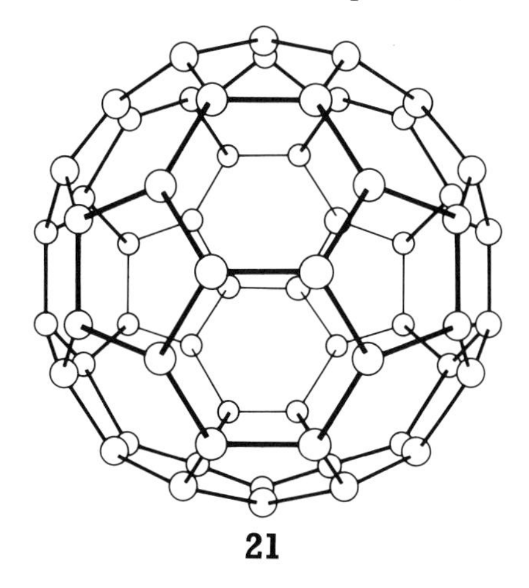
21

that all the rules of classical structural chemistry are satisfied. There is only one way to do this, and that is to place the atoms on the 60 vertices of a truncated icosahedron. The resulting figure has 12 pentagons and 20 hexagons, each pentagon being surrounded by and sharing its edges with five hexagons, each hexagon being surrounded by three pentagons and three other hexagons. It is seen to have a very high symmetry, that of the two regular Platonic solids, the icosahedron and the dodecahedron. It can be imagined as a kind of super benzene, but without hydrogen atoms, or, perhaps better, as a kind of spherical ball of graphite. The hydrocarbon $C_{20}H_{20}$ with the same symmetry (carbon atoms arranged at the vertices of a regular dodecahedron) had been synthesized by the American chemist Leo A. Paquette several years earlier.

The Swiss mathematician Leonhard Euler (1707–1783) found a simple formula relating the numbers of faces (F), vertices (V), and edges (E) of any polyhedron:

$$F + V = E + 2.$$

We can use this to find the number of vertices in any polyhedron that contains exclusively pentagonal and hexagonal faces. Let P be the number of pentagons and H the number of hexagons, then we have:

$$F = P + H$$

$$V = (5P + 6H)/3$$

$$E = (5P + 6H)/2$$

since each vertex shares three polygons and each edge shares two. Applying Euler's formula,

$$(P + H) + (5P + 6H)/3 = (5P + 6H)/2 + 2$$

from which $P = 12$ and H is left undetermined. For $H = 0$, we have the figure with 12 faces, 20 vertices, and 30 edges, the regular dodecahedron. And for each hexagon added, the number of vertices is increased by two. With $H = 20$, we again reach a figure that has the same high symmetry as the regular dodecahedron.

In fact, the same highly symmetric arrangement of 60 protein sub-units is known to occur in the outer coat of several spherical viruses, as was recognized by the molecular biologists Don Caspar and Aaron Klug in 1962, after the fivefold symmetry of these particles had become apparent from diffraction measurements. It is an economical way to cover the surface of the virus particle since the same genetic information can be used to make 60 copies of the protein sub-unit instead of a single giant molecule 60 times as large.

Unfortunately, in our opinion, the C_{60} molecule has been given the name buckminsterfullerene in honor of the American engineer and philosopher R. Buckminster Fuller (1895–1983) who was responsible for the construction of geodesic domes built on the same geometric principle as the molecule. The name is so long and clumsy that it is commonly shortened and degraded to the even less attractive 'buckyball'; the other all-carbon clusters based on 12 pentagonal faces are collectively known as fullerenes. Parents have the privilege of naming their children, but whether the discoverer of a new molecule is entitled to call it anything he likes is another matter; there ought to be rules for deciding this sort of thing.

Today it is possible to answer the question, what does a molecule look like, with relative ease since a whole palette of physical methods is available to determine the atomic arrangement in space as well as the interatomic distances. A direct, visual observation with a microscope is, of course, out of the question, because the wavelength of visible light is much too large for scattering by individual molecules to occur. On the other hand, the wavelength of X-rays is just about the same order of magnitude as the distances between the bonded atoms in molecules. Unfortunately, it is not possible to make lenses that could produce a direct image of a molecule. These missing lenses can be replaced, however, in a certain sense, by a computer, which calculates, from the observed X-ray diffraction pattern, the image that one would obtain in theory with such a system of lenses. The adjacent Figure shows one of the first images calculated in this way by the Scottish chemist John Monteath Robertson (1901–1990); it shows a molecule of the platinum complex of the dyestuff phthalocyanine. This molecule has the formula **22** as can be read, with a little imagination and prior chemical knowledge, together with the atomic coordinates, directly from the X-ray image.

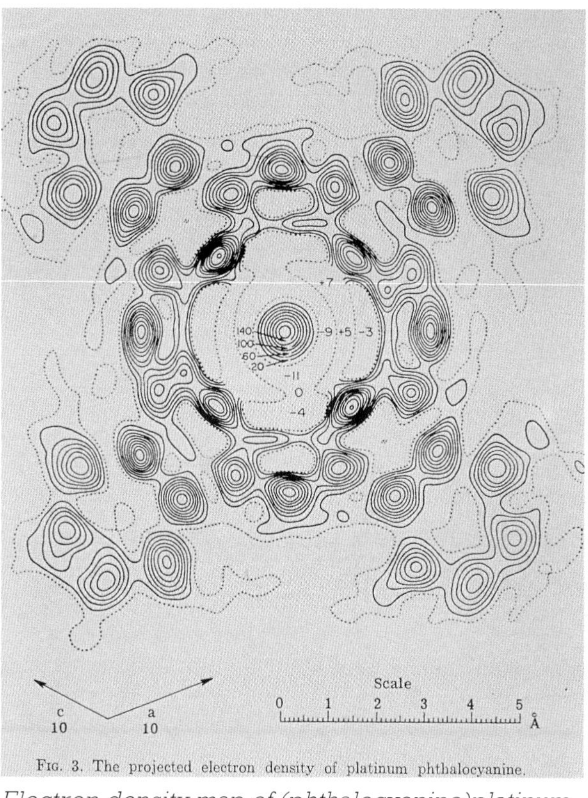

FIG. 3. The projected electron density of platinum phthalocyanine.

Electron density map of (phthalocyanine)platinum, as obtained in an early X-ray diffraction study.

With the help of modern computers coupled with computer-controlled instruments to measure the diffraction patterns of crystals – thousands of separate observations need to be made – such images can nowadays be obtained in a matter of weeks or days, rather than months or even years, as formerly. In this way the atomic arrangements in about 100,000 different molecules have been determined so far by diffraction methods, using X-rays, neutrons, or electrons.

22

Apart from diffraction methods, various spectroscopic methods have played an important part in amassing the vast amount of information about molecular structure that is available today. Most of these techniques depend heavily on symmetry arguments. For example, microwave and infrared spectroscopy, as well as Raman spectroscopy, yield information about the structure of molecules, provided they are not too complex. In the liquid and especially in the gas phase, molecules move with high velocities, collide with one another, tumble about, exchange energy. Depending on the method, one can obtain information about the structure and symmetry of molecules in vibrationally and rotationally excited states, while information about electronically excited states is provided by spectroscopy in the visible and ultraviolet ranges. In nuclear magnetic resonance spectroscopy, each nucleus broadcasts information about its local environment in the molecule; symmetry-related nuclei thus give identical signals.

Returning to benzene, the most conclusive evidence available today for the full hexagonal symmetry of this molecule comes from spectroscopic measurements. These show that the benzene molecule does not oscillate from one Kekulé structure to another but rather vibrates with only small amplitudes about the regular hexagonal structure in which all C–C bonds are of equal length and all C–C–C angles are 120°. This picture is confirmed by the best available theoretical calculations.

When molecules absorb energy and go into excited states, the symmetry of the excited state is not necessarily the same as that of the ground state. For example, if a benzene molecule loses an electron the resulting cation no longer has hexagonal symmetry but jumps around from one lower symmetry structure to another, as shown schematically in the adjacent Figure.

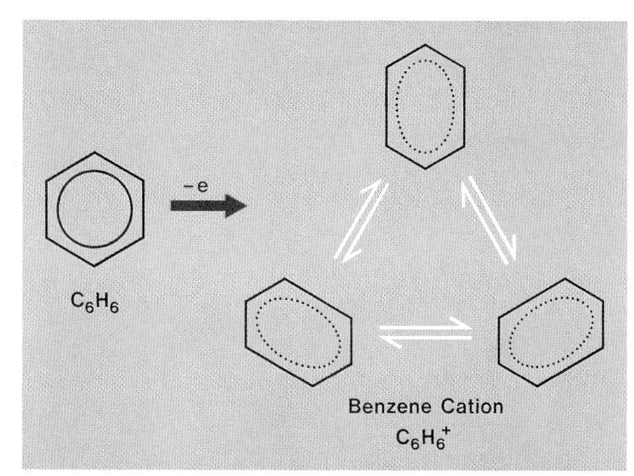

Schematic representation of the loss of symmetry of benzene on ionization.

A curious but surprisingly direct method for obtaining a picture of the spatial arrangement of the atoms in small molecules is provided by the coulomb explosion method[25], shown schematically for the example of the molecule-ion H_3^+, consisting of three hydrogen atoms, as shown in the adjacent Figure. The positively charged molecule is brought to a very high velocity in an accelerator, as used in physics, and shot through an extremely thin carbon foil. During this process, the H_3^+ system does a kind of molecular striptease, whereby its electrons are removed as it passes through the foil. The naked nuclei emerge on the opposite side of the foil with still greater velocity. As these nuclei are positively charged, they repel one another, or in other words, their original structure is expanded. If they are allowed to collide with a detector some distance from the foil, the original structure of the molecule is expanded to such an extent that the impact holes can be observed directly under the microscope. From the positions of the holes, one can draw conclusions about the structure of the original particle, for example, that the hydrogen atoms in H_3^+ are arranged at the corners of an equilateral triangle.

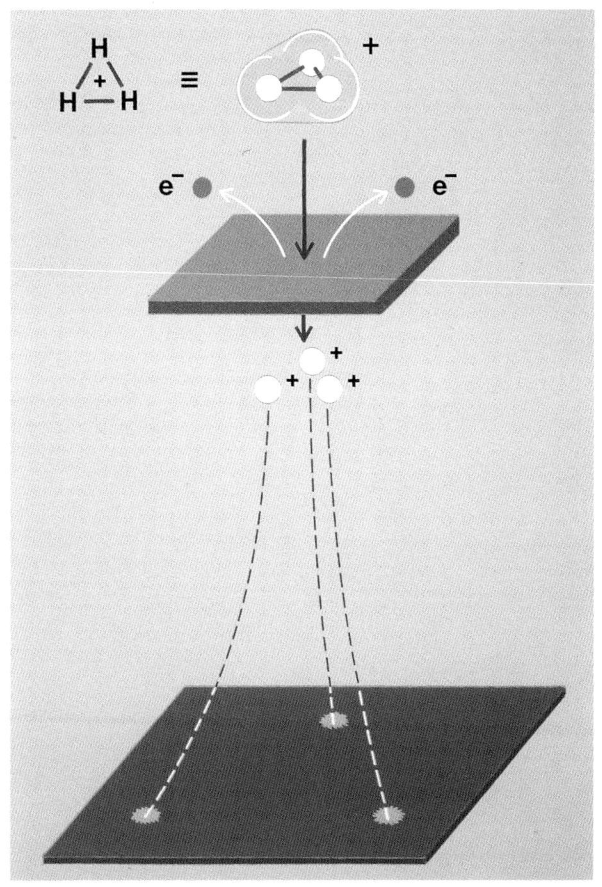

Coulomb explosion method for determining the triangular structure of H_3^+.

VII.

And should not I spare Nineveh,
that great city, wherein are more than sixscore
thousand persons that cannot discern
between their right hand and their left hand;
and also much cattle.

Jonah, Chapter 4

La Droite et la Gauche,
c'est pas du tout la même chose.

Pierre Maurois

The biblical quotation could mean that Nineveh, that great city, contained 120,000 young children, incapable of telling right from left, or perhaps 120,000 inhabitants, regardless of age, who were unable to perform this feat. One of the most fundamental problems of chemistry is concerned with just this concept, with which not only children have difficulty in everyday life.

Everyone knows that the person who gazes at us in the mirror is the wrong way round. He has his watch on his right wrist, his heart beats on the right, and he shaves himself with his left hand, assuming that he is really right-handed. Although up is up and down is down, left and right are interchanged. It may seem puzzling but it is all quite logical. In Chapter V we saw that symmetry operations can be regarded as transformations of the coordinate system. In particular, reflection in the xz plane corresponds to reversal of the direction of the y axis. We go from a right-handed coordinate system (in red) to a left-handed one (in blue), as indicated in the following Figure. The direction of the z axis is unaltered. This is the reason why up and down are not interchanged but right and left are.

Reflection converts a right-handed coordinate system (red) into a left-handed one (blue).

Let us consider again the two-handled vase that we encountered earlier; we know that it possesses, in addition to C_2, the two symmetry operations σ and σ'. In a Gedankenexperiment we now take the mirror-image of the vase from behind the mirror and displace and rotate it until it is brought into coincidence with the original vase. If this can be accomplished, as it can in the case of the vase, then the concepts of 'right' and 'left' become meaningless as far as the vase itself is concerned. The vase and its mirror-image are exactly equivalent.

There are objects, however, whose mirror-images cannot be brought into coincidence with themselves by any dis-

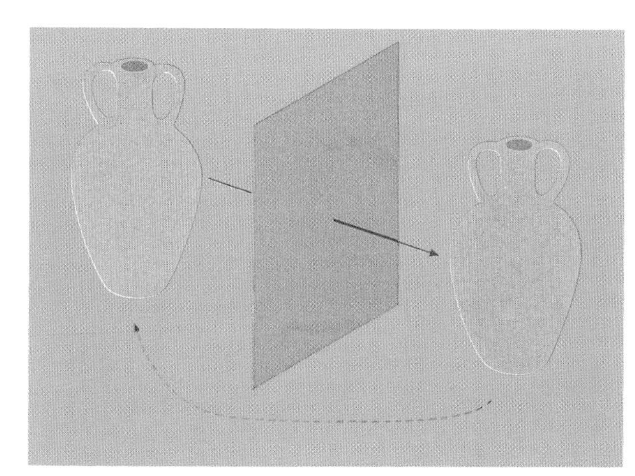

Reflection of a vase.

placement or rotation whatsoever. A hand is the classical example. The mirror-image of a left hand is a right hand (adjacent Figure). A right hand and a left hand cannot be brought into self-coincidence, as everyone knows who has tried to put her right hand into a left glove. An object, or, for us more interesting, a molecule whose mirror-image cannot be brought into coincidence with the original by displacement or rotation is called chiral, from the Greek word for 'hand'. A pair of such molecules is related to each other as left hand to right hand. Rotations and translational displacements are called proper symmetry operations; they convert chiral objects into themselves. Reflection, on the other hand, converts chiral objects into their non-congruent mirror-images and is called an improper symmetry operation. Molecules that contain the symmetry operation σ or a combination of σ with other symmetry operations can be brought into self-coincidence with their mirror images and are hence achiral. Molecules that lack improper symmetry operations and possess only proper ones are necessarily chiral. This makes a convenient test to see whether any given molecule of known structure is chiral or not. Consider the molecule **23**; is it chiral or not?

A pair of hands.

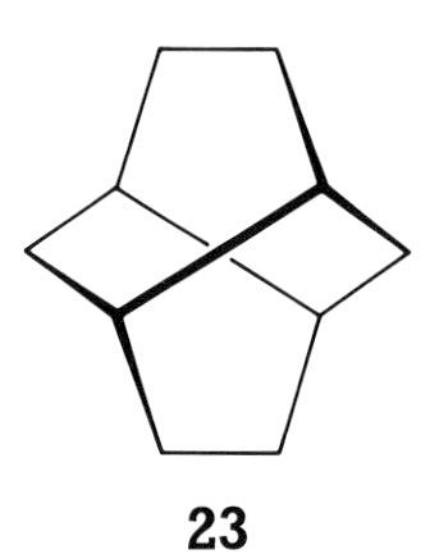

23

One way to answer the question is to make a model of the molecule and also one of its mirror image, and then to move them around in the attempt to recognize whether the two are congruent, that is, whether they can be brought into self-coin-

Van't Hoff's original, three-dimensional models[26].

cidence. It is much simpler to look for the presence of improper symmetry operations in the molecular structure; if none are present then the molecule is chiral, otherwise it is not. Inspection shows that although molecule **23** is highly symmetrical, the only symmetry operations present are twofold rotations – proper symmetry operations, three of them, about mutually perpendicular axes. The molecule is chiral; it has been given the name twistane. In fact, the vast majority of organic molecules, whether they occur naturally in living organisms or have been made synthetically, lack σ operations and are thus chiral. The achiral ones are the exceptions, and they are usually very simple, containing only a few atoms. In the molecular zoo they are really curiosities.

The existence of chiral molecules, related to each other as non-superimposable image and mirror-image, but having otherwise identical structures, was first postulated in 1874 independently by the Dutchman Jacobus Henricus van't Hoff (1852–1911), whose original models are shown in the Figure above, and the Frenchman Achille Le Bel (1847–1930)[13]. They showed that a carbon atom whose four tetrahedrally disposed

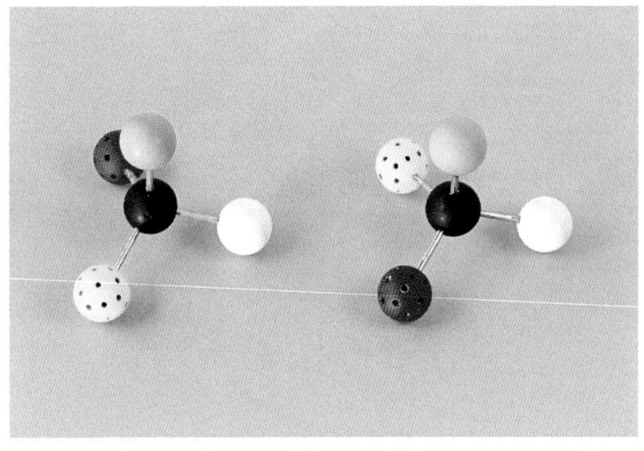

A pair of enantiomorphic models, corresponding to enantiomeric molecules

valencies carry four different atoms or groups (called an 'asymmetric' carbon atom as indicated in the adjacent Figure by four differently colored balls) can exist in two forms, which cannot be brought into coincidence with each other. These two forms are an example of a special kind of isomer, called enantiomers. The one enantiomer is, so to speak, 'left-handed', the other 'right-handed', whereby, as we are not dealing with hands or gloves, we have to leave it open as to which is assigned the one and which the other description. For substances that have been made synthetically from achiral precursors, the enantiomers are present in equal numbers, but for many purposes, e.g. for use in pharmacology, they have to be separated, and this can be a long and difficult matter because the enantiomers do not differ in their chemical or physical properties (except insofar as they interact with chiral environments).

On the other hand, organic substances that occur in Nature, those that are produced by or are building blocks of living organisms, are usually enantiomerically pure, that is, they consist exclusively of molecules of only one sense of chirality; for almost all naturally occurring molecules either the 'left-handed' or the 'right-handed' isomer is present in Nature – not a mixture of the enantiomers.

This property of naturally occurring substances was recognized in an intuitive way by the great French scientist Louis Pasteur (1822–1895) before any detailed theories about molecular structure had been developed. Pasteur made observations on the passage of polarized light beams through solutions of various substances. The electromagnetic waves that constitute a light beam are transverse, that is, the vibration direction is perpendicular to the direction of travel (Figure on the next page). When a light beam passes through certain crystalline substances it is split into two beams that travel with different velocities and are therefore unequally refracted. If

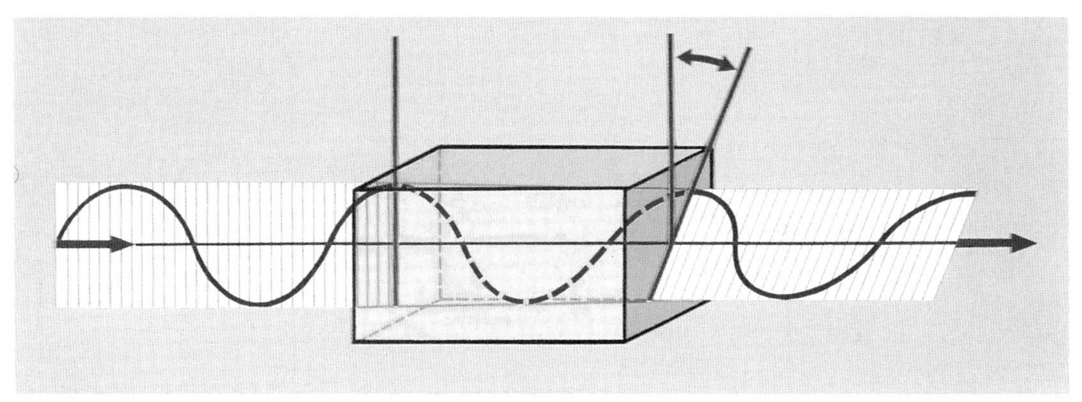

Schematic representation of the rotation of the plane of polarization by an optically active medium.

one of these beams is absorbed (as in a polaroid film) the remaining beam is plane polarized, that is, the vibrations take place only in a single plane. During the first half of the 19th century it became known that solutions of many naturally occurring organic substances have the property of rotating the plane of polarized light – one says that they are optically active.

Tartaric acid is an optically active substance, and, in 1848 Pasteur noticed that crystals of several salts of tartaric acid are hemihedral (this was the word used by crystallographers to describe chiral crystals) – they show a distribution of small facets that is not superimposable with its mirror image (Figure below). It occurred to Pasteur that there might be a general correlation; optically active substances yield hemihedral crystals and vice versa. Indeed, Pasteur formed the idea that "the molecule of tartaric acid, whatever else it might be, is asymmetric and in such a way that the image is not superposable".

But there was a difficulty. One of the minor crystalline products sometimes obtained from wine is the ammonium sodium double salt of tartaric acid. This gave a solution that was not optically active although the crystals were hemihedral, like those of other salts of tartartic acid. Pasteur examined the crystals carefully and made the remarkable observation that some had left-handed facets and some right-handed. He then separated these two kinds of crystal and found that they yielded solutions that rotated the polarization plane in oppo-

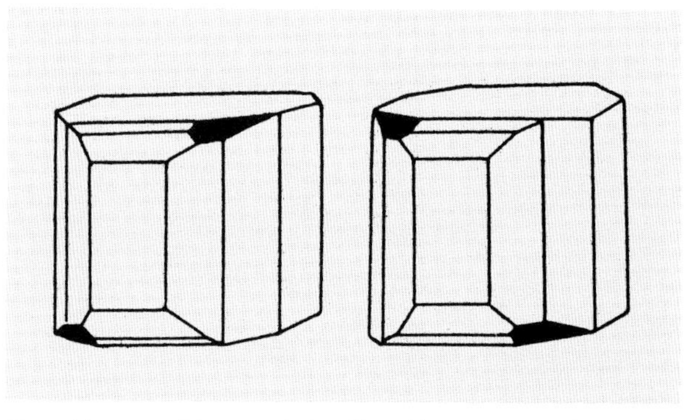

Enantiomorphic crystals of sodium ammonium tartrate[27].

site directions. Indeed, of the two solutions, the dextrorotatory one yielded an acid that was identical with normal tartaric acid, while the other gave a new mirror-image kind of tartaric acid, with properties identical to the normal one except for its opposite sense of rotation. Pasteur saw the connection between the hemihedry of the crystals, the chirality (he called it dissymmetry) of their constituent molecules, and optical activity. What was lacking was a theory of molecular structure, so it was left for van't Hoff and Le Bel to explain these results 25 years later in terms of a more detailed model.

By the middle of this century it was possible by purely chemical methods to correlate the sense of chirality of thousands of compounds. For example, all the 20 amino acids that occur in the polypeptide chains of proteins were known to have the same arrangement of groups, all either the one shown in formula **24** – or all the mirror image of this arrangement. Similarly, the ribose molecules that are part of nucleic acids and the glucose molecules of carbohydrates were all known to be derived from the same basic unit of glyceraldehyde **25** (R = CH_3) – or all from its mirror image. If one model was correct, then so was the other, but from chemical experiments alone it was not possible to decide if both models **24** and **25** are correct as shown below or if both should be reflected in a mirror. It was not known which side of the mirror corresponded to the real world and which to the mirror-image world. Moreover, many chemists believed that this was an unanswerable question.

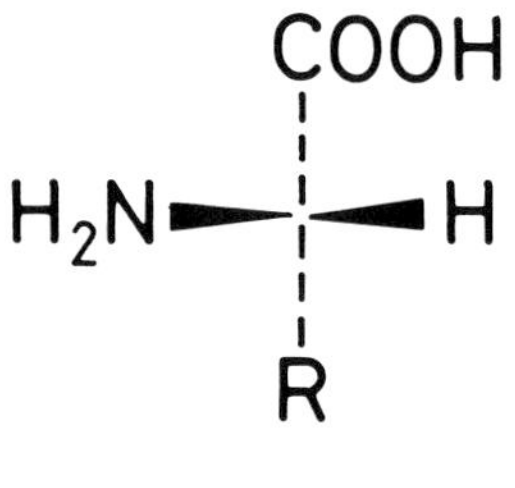

24 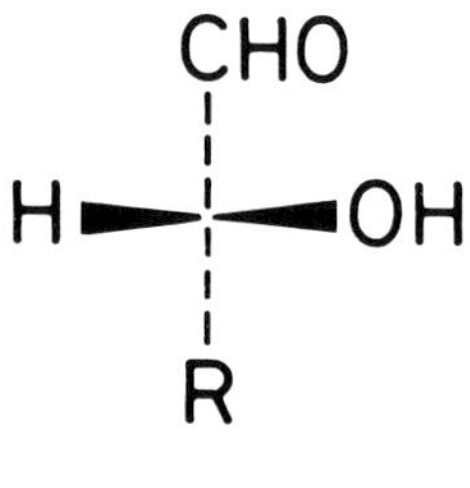 **25** 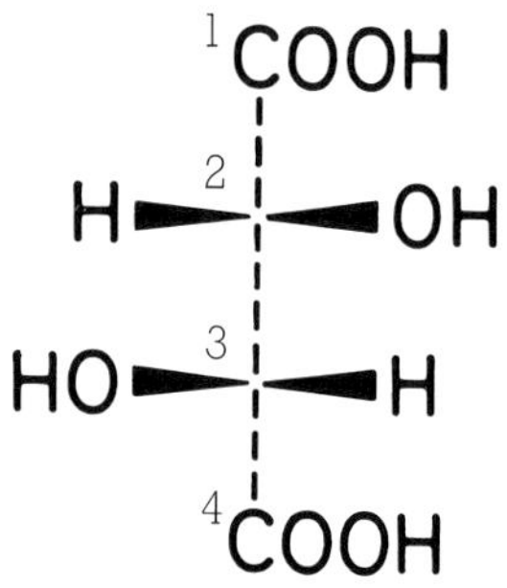 **26**

In view of this uncertainty, the German chemist Emil Fischer (1852–1919) proposed that chemists should agree to use an arbitrary convention to represent one of the two chiral arrangements of atoms at a tetrahedral centre, leaving the question open as to which might ultimately turn out to be correct.

In 1950 the question was answered by the Dutch chemist and crystallographer Johannes Martin Bijvoet (1892–1980), who used a special type of X-ray diffraction experiment to show that the atoms in dextrorotatory tartaric acid were arranged as shown in formula **26** and not in the mirror-image arrangement. The connection between molecular and macroscopic chirality had been made. The method has since been applied to establish the sense of chirality of hundreds of chiral molecules. Barring a few cases where errors were made, the results have been in complete agreement with those based on chemical correlations. Nowadays the 'absolute configuration' of chiral molecules containing tetrahedral centres is usually described in terms of a convention proposed in 1951 by Robert Sidney Cahn (1899–1981), Christopher Kelk Ingold (1893–1970) and Vladimir Prelog.

The CIP system, dubbed after its inventors, consists mainly of a set of rules for putting the four groups around a tetrahedral centre into a priority sequence, a, b, c, d where $a > b > c > d$. The first rule concerns the atomic number of the directly bonded atoms, for example,

$$Br > S > P > O > N > C > B > H.$$

If this is not enough to determine the sequence one proceeds to the second rule, which depends on mass number, for example,

^{3}H (tritium) > ^{2}H (deuterium) > ^{1}H (protium)

or

$^{18}O > ^{17}O > ^{16}O$

If the priority sequence cannot be decided by inspection of the directly bonded atoms, one proceeds outwards along the molecular formula to the next set of atoms and applies these rules as well as other more complicated rules there. Ultimately, for any such 'chiral centre' with tetrahedral valencies, the four groups can be arranged in a priority sequence. For example, in formula **26** for tartaric acid (next page) the groups around the carbon atom labeled 2 are:

a = O(H), b = C(OOH), c = C(H,OH, COOH), d = H

The sense of chirality is then assigned by the following convention: view the chiral center from the direction opposite to the group d of lowest priority. If the sense of rotation of the other three groups, from a to b to c is clockwise, then the chirality sense is designated as *R* (Latin, rectus). If you imagine the three groups to be arranged on the steering wheel of a car, then the rotation a→b→c will steer the car to the right). If the rotation sense is anticlockwise the car will be steered to the left, and the chirality sense is designated as *S* (Latin, sinister).

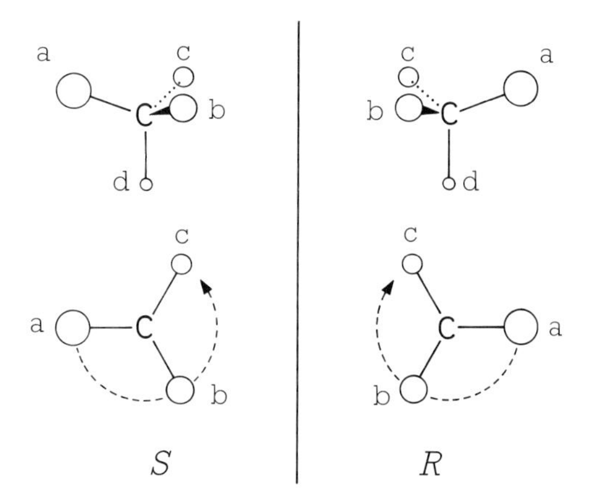

Thus carbon atom 2 in formula **26** is R. Since the two ends of the tartaric acid molecule are related by a C_2 operation, a proper symmetry operation as opposed to an improper one, we can immediately say that carbon atom 3 must also be R. Thus formula **26** represents (*R*,*R*)-tartaric acid, and **27** represents the enantiomeric (*S*,*S*)-tartaric acid.

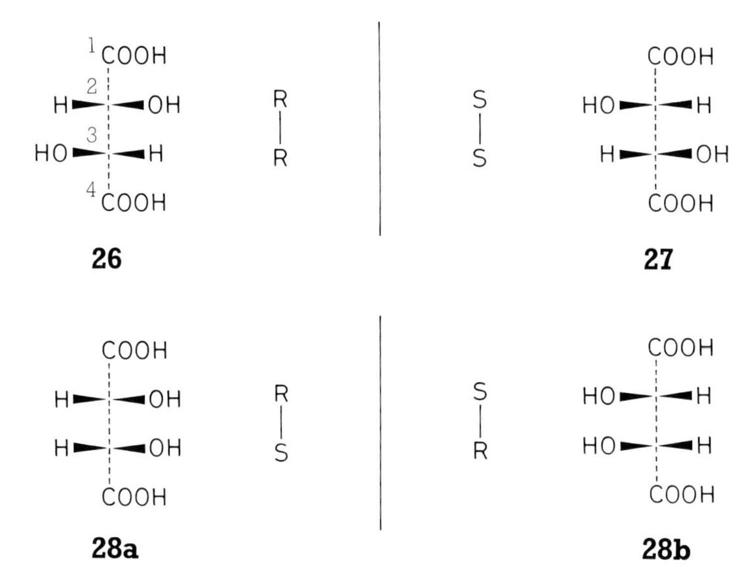

Formulas **28** represent a third isomer, (*R*,*S*)-tartaric acid or *meso*-tartaric acid, in which carbon atoms 2 and 3 have opposite senses of chirality. This molecule is achiral; the atoms can be arranged so that the two halves of the molecule are related by a reflection or an inversion operation. Reflection of **28a** through a mirror converts the *R* group into *S*, the *S* group into *R*, and the result is superimposable with the original. There are thus three isomers of tartaric acid: a pair of enantiomers, (*R*,*R*)- and (*S*,*S*)-, and a *meso*- form, (*R*,*S*).

How many isomers should there be of the structurally related molecule with an additional carbon atom in the chain, trihydroxyglutaric acid, formulas **29**–**32**.

The carbon atoms labeled 2 and 4 are chiral centers since they each have four different attached groups, while carbon 3 does not appear to be a chiral center because two of its at-

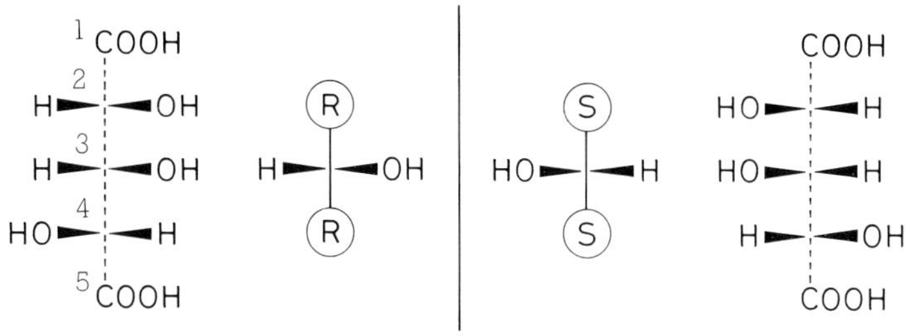

29 **30**

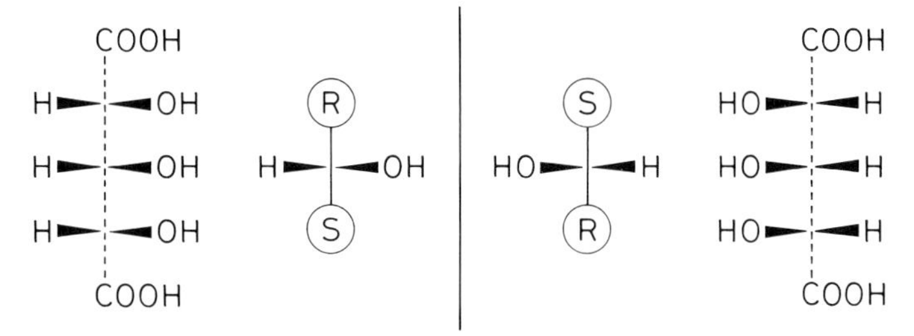

31 **32**

tached groups are the same (–CH(OH)–COOH). As in the previous example, we have the pair of enantiomers, (*R*,*R*) and (*S*,*S*) (**29** and **30**), but what about the *meso*-forms shown in **31** and **32**. Are they the same or are they different? A moment's reflection should reveal that they cannot be superimposed by any series of rotations and translations, and hence they are different. Reflection through a mirror converts each of them into itself and not into the other. So here there are four isomers, a pair of enantiomers (**29** and **30**) and two *meso*-forms (**31** and **32**).

The CIP sequence rules can also be used to order the three groups attached to a central trigonal atom. For example, in the acetaldehyde molecule **33**, where the three groups and the central carbon atom lie in a common plane, the oxygen is a, the methyl group is b, the hydrogen is c. If we view the molecule from one side of the plane, the sense of rotation a→b→c of the three groups is clockwise, from the opposite side it is anticlockwise. In the CIP system, this difference is specified by saying that the one side is the *Re* side, the other the *Si* side. The two sides are related by reflection across the plane and are said to be enantiotopic.

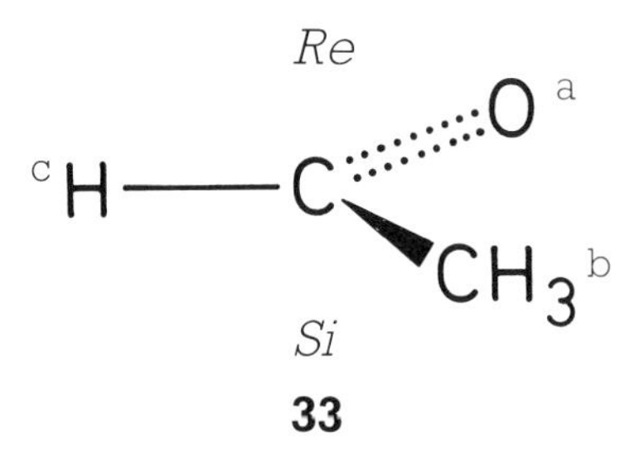

Thus we can distinguish between the two faces of a molecule that contains a plane of symmetry as its only symmetry element. Why should one ever want to make such a distinction? It turns out that in many biological reactions the two enantiotopic faces of a non-chiral molecule like acetaldehyde behave quite differently; a reaction can occur specifically on one face but not on the other. This may appear puzzling at first sight but an analogy from everyday life may help to make things clearer. A teacup has a plane of symmetry as its only symmetry element. The two sides of the cup are enantiotopic, and a right-hander will always drink from the same side. After many years of service, one side of the cup may be quite worn while the other is still practically unused, as shown in the adjacent Figure.

Teacup after many years of use by one of the authors.

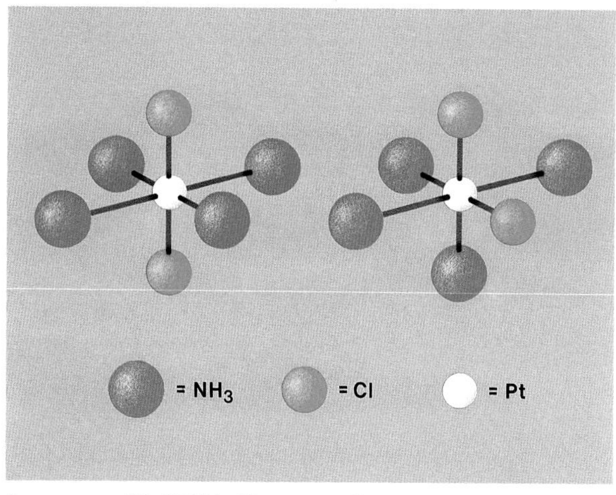

Isomers of $Pt(NH_3)_4Cl_2$ according to Werner.

For quite some time it was believed that optical activity was a property of molecules containing carbon and necessarily connected with the presence of 'asymmetric carbon atoms', those with four different groups attached by tetrahedrally arranged valencies, as proposed by van't Hoff and Le Bel. This belief was shattered by Alfred Werner (1866–1919), who showed that a whole world of molecules, so-called coordination complexes, was based on the arrangement of six ligand atoms about a central atom with octahedral geometry, leading to new types of isomers[28]. For example, there are two substances with the molecular formula $Pt(NH_3)_4Cl_2$, one with the chlorine atoms at adjacent vertices of an octahedron, the other with them at opposite vertices (Figure above). Clearly, quite new types of complication can occur, and symmetry can be helpful in sorting them out.

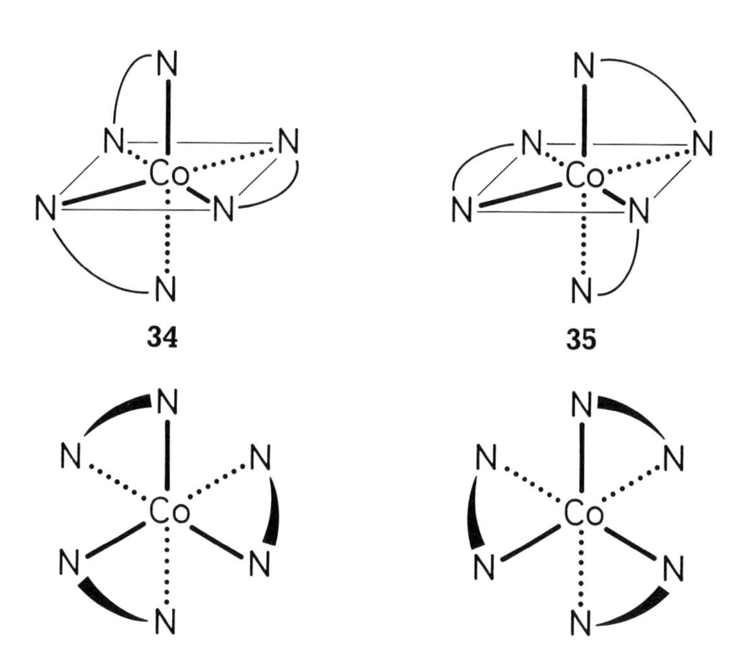

The molecule ethylenediamine $H_2N–CH_2–CH_2–NH_2$ (en for short) can be joined to a central metal atom through both of its nitrogen atoms, forming a closed five-membered ring, known as a 'chelate' ring (Greek, chele, a claw). The results of attaching three such chelate rings to a cobalt atom are shown in formulas **34** and **35**. The two mirror-image figures **34** and **35** can be seen to contain only C_3 and C_2 symmetry operations, i.e., proper symmetry operations; in other words the two figures are chiral and therefore represent enantiomers. In one figure, all three chelate rings are turned as in a right-handed screw, in the other as in a left-handed screw. Indeed, Werner separated the substance $Co(en)_3Cl_3$ into optically active forms, and their 'absolute configuration' has since been established by X-ray structure analysis of the crystals by the Bijvoet method.

One of the most important and best known chiral molecules, in view of its central role as information carrier in molecular biology, is deoxyribonucleic acid (DNA), the famous double-helix, whose structure was deciphered in 1953 by James D. Watson and Francis H. C. Crick. As can be seen from the adjacent Figure, it consists of two cross-linked, right-handed helices (the same sense of helicity as a normal screw, the prototype of a chiral object)[29].

The transformations of molecules in living organisms take place within giant organic

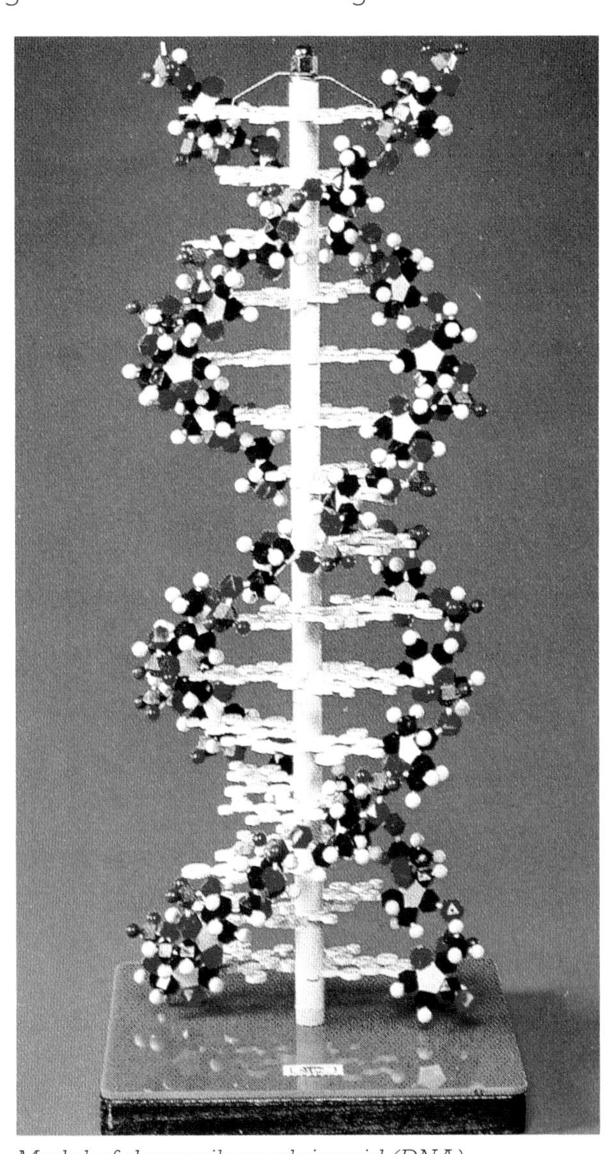

Model of deoxyribonucleic acid (DNA).

bio-molecules known as enzymes. A typical example is lysozyme, discovered in 1922 by the same Alexander Fleming (1881–1955) to whom we are indebted for the discovery of penicillin. The structure of this macromolecule was determined by David Chiltern Phillips and his collaborators by X-ray analysis in 1965. The accompanying Figure gives an impression of the complexity of this chiral system.

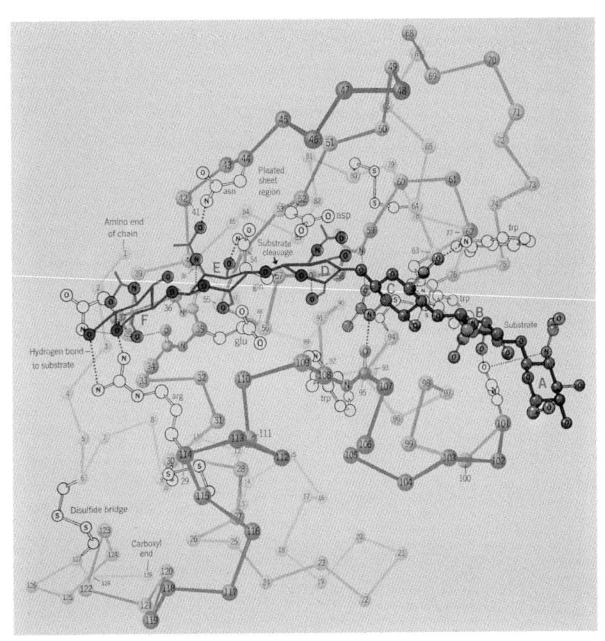

Model of lysozyme with guest molecule[30].

Nature thus manipulates chiral molecules with the help of chiral enzymes that contain active sites, cavities within which the actual chemical reactions take place. In the Figure above, such a guest molecule is indicated in darker tone within the lysozyme host molecule, which seems to grasp it almost like a hand. Since the molecule to be altered as well as the enzyme itself is chiral, it is obvious that the sense of chirality of the two must be matched. Just as a left-hander has difficulty with a pair of scissors, which is designed for the right hand and thus cannot be grasped properly by a left hand, an enzyme is not very successful in catalyzing a reaction of a molecule with the wrong sense of chirality. In the next Figure this is illustrated in a somewhat naive manner. The two mirror-image models on the left represent 'left' and 'right' enantiomeric molecules of phenylalanine. Only one of these, say the 'right', sits comfortably and can then be subjected preferentially to some alteration or other. When, on the molecular scale, the hand is replaced by an enzyme, this preference can be so strong that only one enantiomer, say the 'right' one, will be bound to the enzyme and then altered. If the other enantiomer is presented nothing happens to it.

Similarly, two enantiotopic groups of a mirror-symmetric molecule may suffer different fates in a series of biological reactions where enzymes are involved. In the achiral

Naive explanation of the specificity of an enzyme for one enantiomer over the other.

molecule aminomalonic acid, **36**, the two COOH groups may appear to be equivalent, but when the compound is presented to bacteria, one of these groups, say 1, is split off as carbon dioxide, leaving a molecule of glycine containing specifically the COOH group 2. The spatial arrangement of C(1)OOH, NH_2, and H is the same as in the naturally occuring series of amino acids (compare **24**).The arrangement of C(3)OOH, NH_2, and H is the mirror image. Thus the two ends are related as a right hand to a left hand, and the chiral enzymes have no difficulty in distinguishing one from the other.

36

We are now in a position to understand why the molecules of most compounds involved in living processes occur in only one sense of chirality. The presence of both enantiomers would require two sets of enantiomeric enzymes, which would mean at least a duplication of the metabolic energy required to build the essential chemical machinery and additional genetic information. However, a few enzymes have been identified that can handle both enantiomers and even interconvert them. For example, mandelate racemase interconverts the enantiomers of mandelic acid, which is produced in the biochemical pathway in one enantiomeric form and utilized in the other. Since the equilibrium constant for this reaction must be unity, the rates of the forward and backward reactions must be equal at equilibrium. So it seems that here we have an enzyme that can cope with 'right' and 'left' hands with the same efficiency.

In *Through the Looking Glass* Lewis Carroll describes how Alice slips through the mirror into another world of the opposite chirality. Is it a coincidence that this book, which makes several references to the left/right dichotomy, was first published in 1872, only two years before the papers of van't Hoff and Le Bel about the explanation of optical antipodes in terms of spatial molecular models? Lewis Carroll was the pseudonym of Charles Lutwidge Dodgson (1832–1898), a mathematician at Christ Church college at Oxford, who may well have been in a position to learn of the new exciting discoveries in chemistry that were calling for clarification in structural terms. One of Dodgson's closest friends was the chemist Augustus George Vernon Harcourt (1834–1919), Fellow of the same Oxford college and one of the first to study the rates of chemical reactions. Had Harcourt told Dodgson about the recent puzzling findings of Johannes Wislicenus (1835–1902) concerning lactic acid? Wislicenus had shown that one of the substances present in muscle appeared to be identical with lactic acid obtained by fermentation of milk, except that solutions of the two substances rotated plane polarized light in opposite senses, and he was aware that this result was incompatible with the then current structural theory and called for special explanation in geometrical terms. In 1873 he wrote "The facts force an explanation of the difference between

Alice, as seen from the two sides of the mirror, according to Tenniel's illustrations.

isomeric molecules with the same structural formula in terms of different arrangements of the atoms in space." Indeed, as van't Hoff acknowledged, it was this comment that stimulated him to occupy himself with the problem of the spatial arrangement of groups attached to carbon atoms. One might speculate that Dodgson and Harcourt may have discussed this stereochemical problem, perhaps at high table dinner, and that this may have been one of the underlying and perhaps even unconscious stimuli for *Through the Looking Glass.*

However, if we can believe Tenniel's illustrations (Figure above), Alice herself is curiously left unchanged as she passes through the mirror; her right side remains right and her left remains left, so that we may conclude that the spatial arrangement of all the molecules she consists of is also left unchanged[31], If that were really the case, then one can only feel sorry for poor Alice! The rest of the world behind the mirror surface must certainly be reflected, and accordingly all the molecules and fundamental particles of which she consists, referring to the unchanged Alice, have the opposite, wrong sense of chirality. Before she passes through the looking glass, Alice even poses the question in her monologue with Kitty, her cat: "Perhaps Looking-glass milk isn't good to drink?''. Alice's unchanged enzymes would be unable to assimilate and digest the molecules of the reflected food, so that the poor child would die of hunger, quite apart from the fact that her whole metabolism would presumably have gone out of joint even earlier. It might even be worse! According to physics the encounter between the cis-specular Alice and the reflected matter on the other side might lead to a loud bang and to the total transformation of our heroine into radiant energy.

Until 1956 most scientists took it more or less for granted that the laws of physics are mirror-symmetric; this was certainly true for the classical gravitational and electromagnetic interactions and it was generally assumed to hold for all other possible interactions as well. This belief was even elevated to the status of a principle – the principle of the conservation of parity, as it was called. In 1956, however, this was questioned by two young Chinese-American physicists, Chen Ning Yang and Tsung Dao Lee, who suggested that parity may not hold in the world of elementary particles. In fact, they predicted that parity might be violated in the disintegration of certain particles, including the β-decay of certain radioactive nuclei (in β-decay the radioactive nucleus emits an electron, thus raising its atomic number by one). An experimental verification soon followed; electrons emitted from a sample of a β-emitter were found to be predominantly left-handed (each electron is associated with a 'spin', and the sense of spin was found to be preferentially related to the direction in which the electrons were propagated as in a left-handed screw, as shown in the Figure below). This means that parity is violated at this level of physics, the level of the so-called weak interactions.

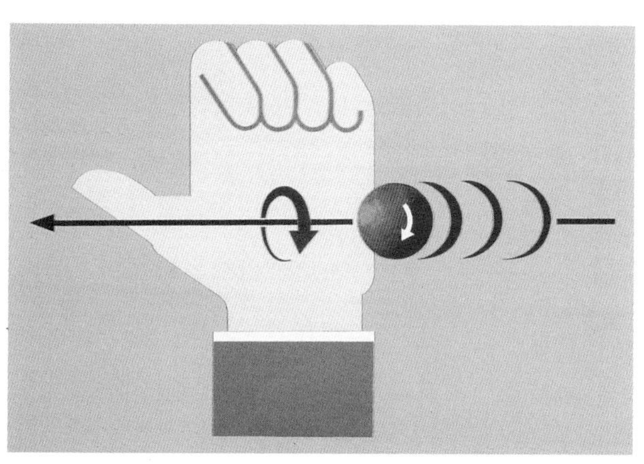

Electron, propagated as in a left-handed screw.

Most physicists were astounded by the news. The reaction of Wolfgang Pauli (1900–1958) is well known. When he learned that the crucial experiment was about to be made he commented: ''I do not believe that the Lord is a weak left-hander, and I am ready to bet a very large sum that the experiments will give symmetric results''. Pauli had a very deep feeling for symmetry in physics, but for once he was wrong. On the other hand, one physicist who was not surprised by the news was Paul Adrian Maurice Dirac (1902–1984) who had predicted the existence of anti-matter some twenty years earlier. In 1949 Dirac had written: ''I do not believe that there is any reason for physical laws to be invariant under reflections, although all the exact laws of physics so far known have this invariance''[32]. Dirac went on to explain that the laws of physics must be unaltered by rotations and translations, since these can be generated by infinitesimal changes – but this does not apply to reflections. Here the fundamental difference between the two kinds of symmetry operations, rotations and reflections, between proper and improper symmetry operations, is expressed in the clearest possible way, and one can only be astonished at the directness and simplicity of Dirac's argument.

The downfall of parity at the weak interaction level leaves the door open for a possible explanation of the origin of biomolecular chirality. An atom can no longer be regarded as having spherical symmetry. The weak interaction makes a pair of mirror-image molecules slightly inequivalent as far as their energies are concerned. One is slightly more stable than the other, and from calculations it has been claimed that the energetically preferred ones are just those which occur in the bio-molecules of our earth – the L-amino acids and the D-sugars. However, the calculated difference is extremely small; it corresponds to an excess of one molecule of the energetically favored species in about 10^{18} molecules of the mixture at normal temperature. Remember that if you toss up 10^{18} coins you do not expect to get exactly half heads and half tails. In fact, the odds that this should happen are infinitesimally small. From statistical theory there is about a one in three probability that the deviation from exact equality will exceed the standard deviation of the sample, which is here the square root of the total number of trials, in this case 10^9. So the preference due to the asymmetry of the weak interaction might appear to pale into total insignificance, were it not for the fact that the evolution of biological systems has taken place over a very long time period and has involved many orders of magnitude more molecules than 10^{18}. Of course, other explanations are also possible. If life indeed happened only once then it was almost certainly on one side of the mirror plane or the other, not on both. It could have been merely a matter of chance. The question is still undecided.

Alice's journey into the reflected world raises, however, a very practical problem that has a rather surprising solution. Chemists have learned how to make 'left' molecules out of 'right' ones in the laboratory, or, in other words, à la Alice, how to pass chiral molecules through the mirror. How is this done? A preliminary comment is called for.

Up till now, for example in the discussion of the substitution reactions of benzene, we have considered exclusively the structure and symmetry of the molecules of the starting material , the so-called educt, and of the product, without bothering about the way in which the transformation actually takes place. In this respect, however, it should still be possible to show with the aid of our simple models how the reacting molecules approach one another, alter the relative positions of their atoms, and then move apart again. This process, which one must imagine spatially and in relationship to models, is called the reaction mechanism. One of the first to develop such ideas was the already mentioned Marc Antoine Augustus Gaudin, who was in this respect, however, far ahead of his time. In the next Figure, we show, as an example, how he imagined the transformation of a trigonal bipyramidal molecule with a C_3 axis to a square pyramidal product with a C_4 axis; in this case, according to our convention, the product has a higher symmetry than the educt.

Now we return to the problem at hand, the formulation in terms of our model of the mechanism of the 'mirror-image' reaction. The simplest chiral molecule imaginable consists of four different atoms that do not lie in a common plane. In the

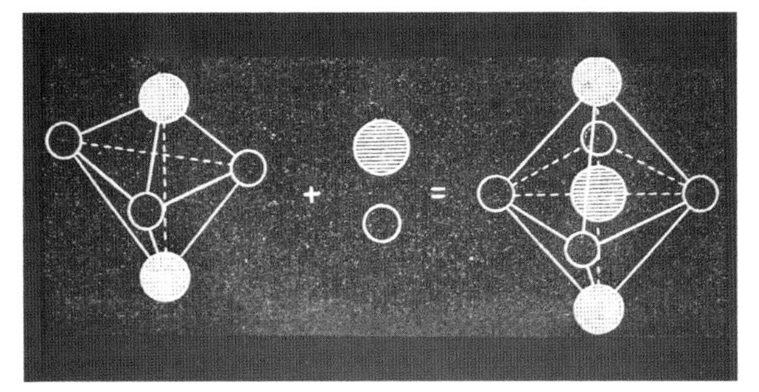
Gaudin's imagined reaction mechanism.

Figure below, they occupy the corners of a regular tetrahedron, but this is only for aesthetic reasons.

If one wishes to convert the model on the left into its mirror-image on the right, one has to move one of the four atoms somehow or other (a few arbitrarily chosen ways are indicated in the left picture) through the plane containing the other three atoms. At the moment when the selected atom actually passes through this plane – for example, at one of the three spots indicated by small circles – the entire molecule is in a planar transition state, since at this instant all four atoms lie in a common plane. As a result, in the transition state on the way from 'left' to 'right' our molecule possesses the reflection σ as a symmetry operation, the mirror-plane coinciding with the common plane of the atoms. Presumably, the reader will find nothing surprising about this, since intuition may well suggest that such a mirror-symmetric arrangement of the atoms must necessarily arise in the transformation of a 'left' into a 'right' molecule by continuous deformation. This is by no means the case, however, as we demonstrate with a simple example.

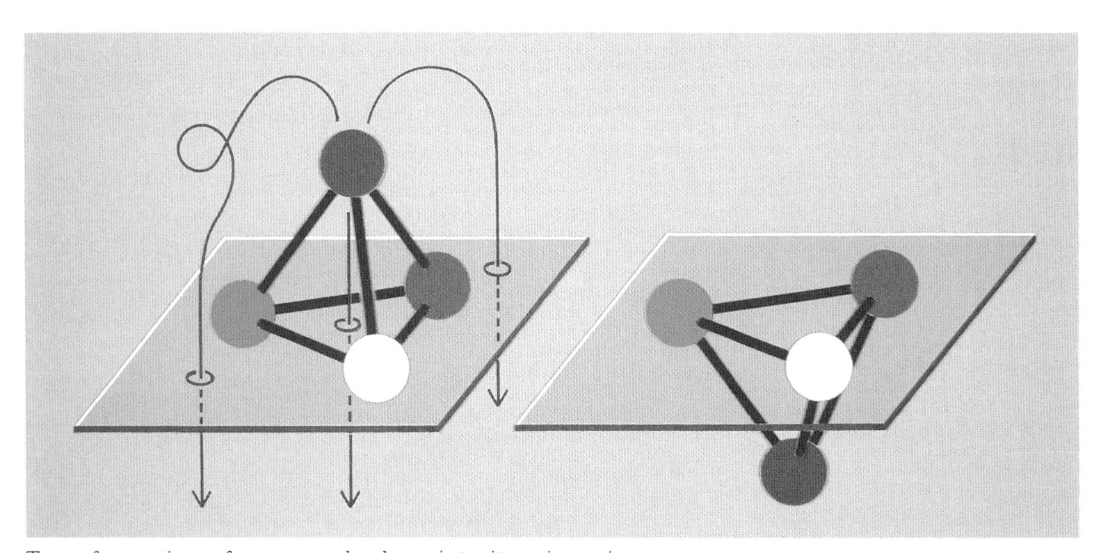
Transformation of one tetrahedron into its mirror image.

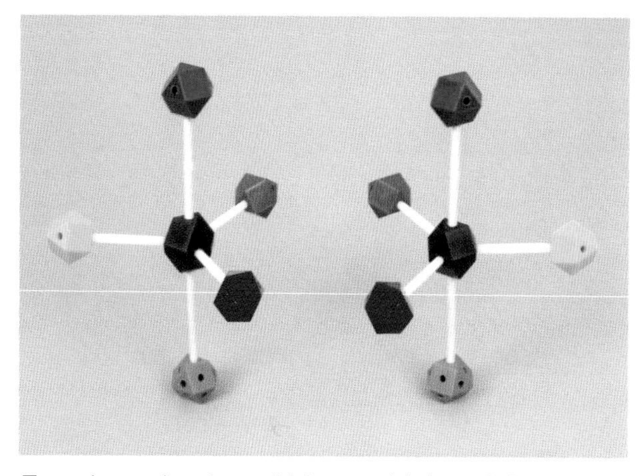

Enantiomeric trigonal bipyramidal models.

The adjacent Figure shows the model of a molecule in which a central atom – in practice it is usually phosphorus – is linked to five other atoms, all different, at the vertices of a trigonal bipyramid. If the three equatorial atoms were the same, the molecule would have the symmetry operation C_3 along the vertical axis as well as three vertical reflection planes. With the five vertices occupied by different atoms, the molecule has no symmetry. The model on the right is easily seen to be the mirror image of the one on the left. How can one pass from one to the other?

One possible process (it is probably not the way it actually goes but it illustrates the general idea) involves interchange of an equatorial atom and an axial one. This could happen, for example, by rotation of the pair around an imaginary axis joining their mid-point to the central atom of the bipyramid (adjacent Figure). It is seen that a sequence of three successive interconversions of this type is sufficient to transform the educt in the top left corner into the mirror-image product in the top right corner. The process that actually occurs is probably a little more complicated. In it two of the equatorial atoms become axial while the initially axial pair becomes equatorial, one initially equatorial atom remaining unmoved – the pivot atom. Again, it needs a sequence of three such interchanges to convert a chiral trigonal bipyramidal molecule into its enantiomer.

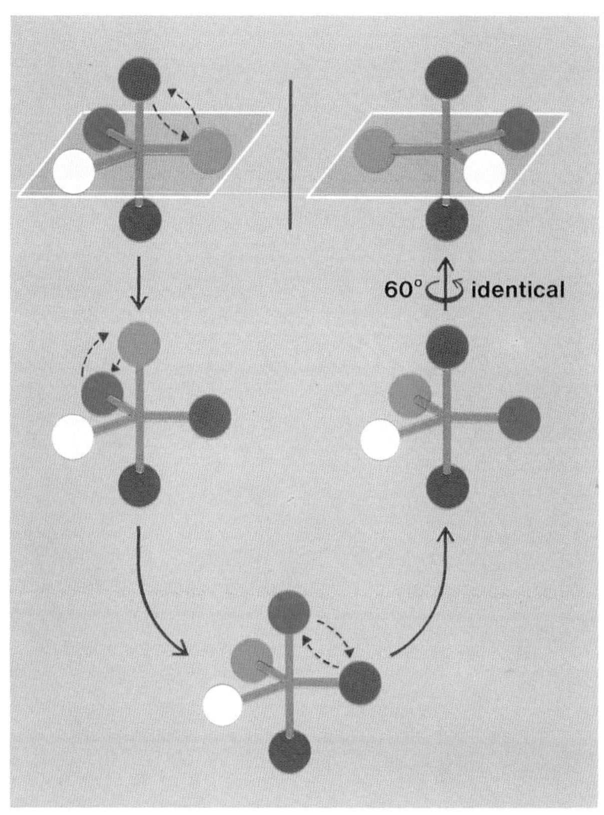

Possible sequence of steps involved in the transformation of a chiral trigonal bipyramid into its enantiomer.

What is remarkable and perhaps even surprising is that at no point in these continuous transformations does an arrangement of the six atoms occur which possesses the reflection σ as symmetry operation! In other words: we have done the trick of moving from the front side of the mirror to the rear side without ever passing through the mirror itself! In a manner contrary to intuition, we have, so to speak, gone round the mirror into the mirror-image world. Perhaps most astonishing is that this is the way preferred by Nature, for, as far as we know today, the transformation of 'right' into 'left' molecules occurs by such, for us perhaps strange, mechanisms.

lichtung

manche meinen
lechts und rinks
kann man nicht
velwechsern
werch ein illtum !

Ernst Jandl[33]

VIII.

As I was going up the stair
I met a man who wasn't there.
He wasn't there again today,
I wish that man would go away.

Anonymous

To help the primarily intuitive concept of symmetry to be grasped more precisely, we have introduced the idea of symmetry operations (σ, C_2, ...). We recapitulate: a symmetry operation transforms an object, for example, a molecular model, 'into itself'. By this we understand that after the symmetry operation has been carried out, we cannot distinguish the final state of the object from its initial one. As long as we consider only idealized objects, like the trigonal bipyramid or other geometric figures, or as long as the symmetry operations are carried out purely in the spirit of a Gedankenexperiment, there are no difficulties. This applies to all the cases we have discussed so far. They all concern molecular models that we have implicitly taken to be rigid and perfectly symmetric. In the previous chapter, as far as symmetry was concerned, we discussed only rigid models of educts, products, and intermediate states, whereby in the last case we had to imagine that we stopped the motions of the atoms corresponding to the assumed reaction mechanism at some suitable point. In reality it is all somewhat more complicated. One reason for this can be understood if we try to answer the question, how, and under what circumstances, can the indistinguishability of the initial and final states be established?

A quite simple example shows that the answer to the question, whether an operation is to be considered as a symmetry operation or not, depends on the degree of 'depth' of the observation. A human being shows superficial bilateral symmetry. In our language, this means that reflection across the plane S', as shown in the left side of the Figure on the next page, is a symmetry operation σ. However, if we look as an

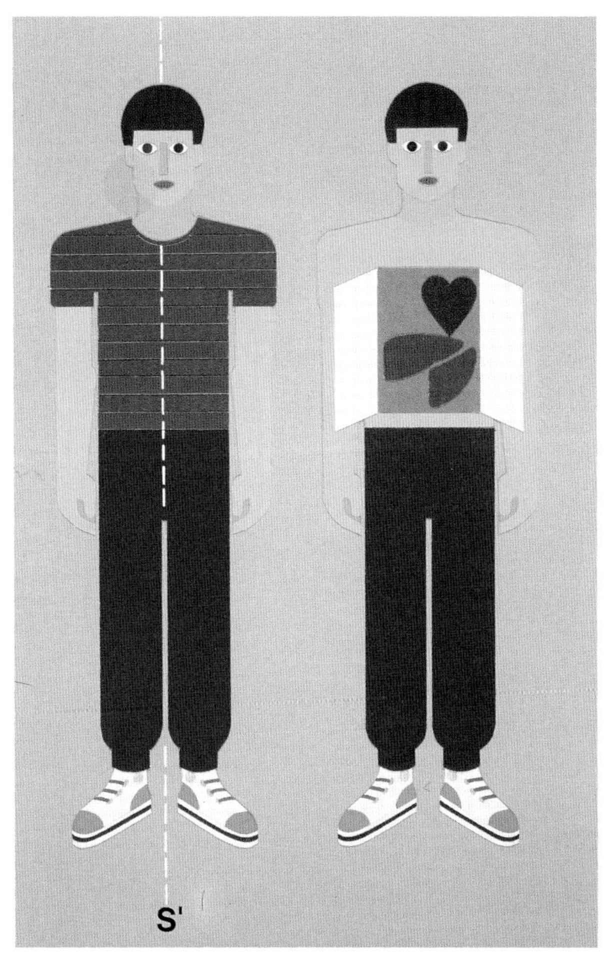

The apparent symmetry of the boy is only superficial.

anatomist into the innards of the creature the same reflection is no longer a symmetry operation at this level, since the heart and other organs are placed completely unsymmetrically.

Also the five-pointed starfish which appears to have the symmetry operation C_5 (= rotation of 360°/5 = 72°) only has this property when we look at it from the outside. A zoologist would explain to us that the starfish shows at most bilateral symmetry, that is, mirror symmetry, when it is dissected, and even less at the molecular level.

Whether a rotation or a reflection is a symmetry operation or not obviously depends, in these two examples, on how one makes the comparison between the initial situation and the final situation, that is, the result of the operation. It can well be the case that we have no control over the 'how'. Here is a simplified example. In the upper left part of the following Figure we see a vase with some pattern burned into it. Rotation by 180° produces the situation shown on the right, where the pattern is now turned away from the viewer.

Does this 180° rotation correspond to a C_2 symmetry operation or not? This depends on whether we make our observation in the light or in the dark. In the first case, the answer is: no! The final situation differs quite clearly from the initial one, as anyone can see. In the dark, however, the vase is invisible,

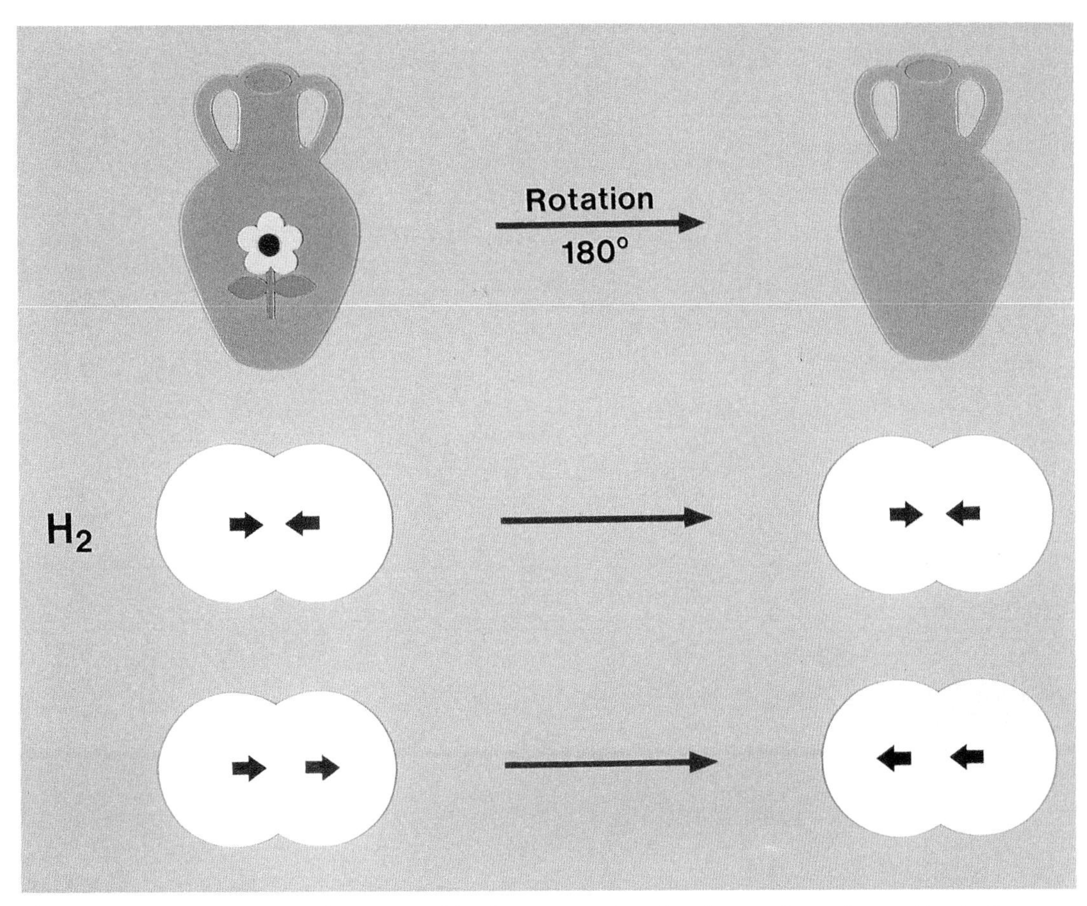

The apparent symmetry of vases and hydrogen molecules depends on the method of observation.

and we have to rely on tactile stimuli. The two situations cannot be distinguished by touch alone; under these conditions the 180° rotation has become a symmetry operation. One might note, moreover, that for a blind person the 180° rotation will always be a symmetry operation for the vase in question, irrespective of whether it is light or dark.

Could similar effects also play a part in observations of molecules? Yes, definitely, as one can demonstrate even with the simplest of all molecules, the diatomic hydrogen molecule H_2. The electron cloud of this molecule has a high symmetry. In particular, a rotation of 180°, which exchanges the two nuclei, brings the molecule into self-coincidence when we picture it with atomic models, for example. Since only the electron cloud is important for its chemical properties, this means that, as far as the chemist is concerned, the H_2 molecule is a highly symmetric object, which shows, in particular, the symmetry operation C_2. One knows, however, that the nucleus of a hydrogen atom behaves as a tiny magnet; in the previous Figure the north pole of this magnet is represented as an arrowhead. When two hydrogen atoms form a hydrogen molecule, they can do so with the two nuclear magnets oriented either oppositely or in the same direction. The first kind of molecule is called para-hydrogen, the second ortho-hydrogen. From the last Figure, it is obvious that when we include the orientation of the nuclear magnets in our consideration, rotation of 180° is a symmetry operation only in the first case; in the second case, in ortho-hydrogen, the direction of both nuclear magnets is reversed by this rotation and hence the initial and final situations are distinguishable, at least in an external magnetic field. If one utilizes an observational method that can register the direction of the nuclear magnets, then ortho-hydrogen will appear to have a lower symmetry than para-hydrogen.

We thus have here on the molecular plane a very similar situation to that of our vase. For the 'blind' chemist, who can touch only the electron cloud, hydrogen is a highly symmetrical molecule, whose symmetry is expressed in its reactions with other molecules. From the viewpoint of the physicist or physical chemist, who can 'see' also the magnetic properties, there are two kinds of hydrogen molecules; one kind with the high symmetry favored by the chemist, and another kind, about three times more frequent, with lower symmetry. It may be remarked by the way that such a difference in symmetry which depends of the manner of observation can be very appealing from an aesthetic standpoint. Thus the regular platonic solids decorated with Escher motifs (next Figure) still have their full symmetry from a tactile point of view, but this is reduced visually by the lower symmetry patterns. Few will be able to resist the fascination that emanates from this irritating discrepancy.

Another factor, critical for symmetry considerations in chemistry, which we have ignored so far in our discussion, is the time, or rather the time interval in which we can make our observations. Here again is another simple example in which

Platonic solids decorated with Escher patterns, designed by Doris Schattschneider and Wallace Walker[34].

we once more make use of our vase. Let us assume that the vase rotates with constant angular velocity about its vertical axis; this can be arranged by placing it at the centre of a gramophone turntable. Our observation instrument is a photographic camera. The next Figure shows the sort of picture we would obtain with a flashlight photograph or, at the other extreme, with such a long time-exposure that the vase makes several complete rotations in the interval.

We see that if the time-scale of the observation is very short relative to the time needed for a revolution of the vase, then only the reflection in the plane through the middle of the vase will be registered: under flashlight conditions, the vase has only this lower symmetry. On the other hand, if the time-scale of the observation is very long compared with the period of rotation then we obtain a different result. All positions average out, and the photograph corresponds to a rotationally symmetric vase, that is, one with an apparently much higher symmetry. This kind of differential symmetry assessment accord-

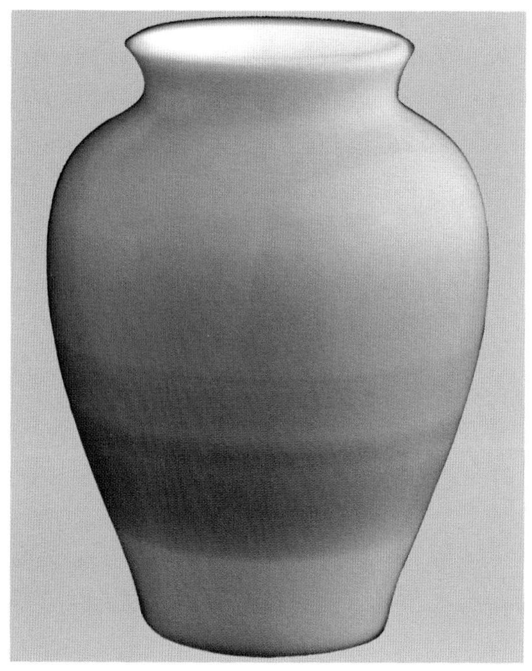

Flashlight and time-exposure photographs of a rotating vase.

ing to the time-scale of the method of observation is very important in chemistry. Here is an example.

In the previous chapter we mentioned that cyclohexane does not exist as a planar hexagonal ring skeleton, as Baeyer had earlier supposed, but rather – as Sachse had maintained – as a non-planar form. The next Figure shows a molecular model of the chair form; the twelve hydrogen atoms are divided into two groups and characterized with different colors (white and red). The white hydrogen atoms stand alternatively above and below the carbon atoms, their C–H bonds being parallel to the molecular C_3 axis, while the red hydrogen atoms occupy positions near the equator, the mean plane of the ring. They are therefore known as 'axial' and 'equatorial' hydrogen atoms.

The cyclohexane molecule has the ability to undergo a ring inversion process in which the carbon atoms carry out the indicated motions (for the sake of clarity, hydrogen atoms have been omitted in **37**), as can be readily demonstrated with molecular models.

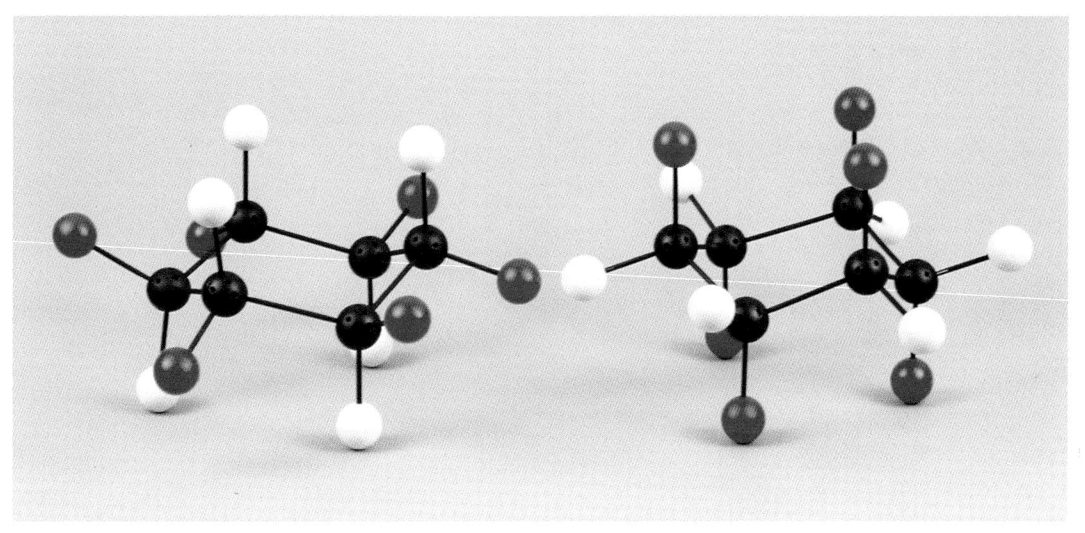

The axial and equatorial hydrogen atoms of cyclohexane are interchanged by ring inversion.

37

The result of this ring inversion is to interchange the orientations of the two sets of hydrogen atoms. The model on the left of the Figure is converted into the model on the right, in which the initially axial hydrogen atoms (white) have become equatorial, and the initially equatorial ones (red) have become axial. This is the inversion process that Mohr suggested as being responsible for the failure to isolate more than one isomer of a mono-substituted cyclohexane derivative. Molecules with the substituent in an axial position and those with the substituent in an equatorial position would be in rapid dynamic equilibrium with one another so that only the more stable of the two isomers would be isolable.

A physical method, nuclear resonance spectroscopy, permits hydrogen atoms in different environments to be distinguished from one another. Each kind of hydrogen atom produces a sharp peak signal, whose position on a recording strip is characteristic for the environment of the atom and whose

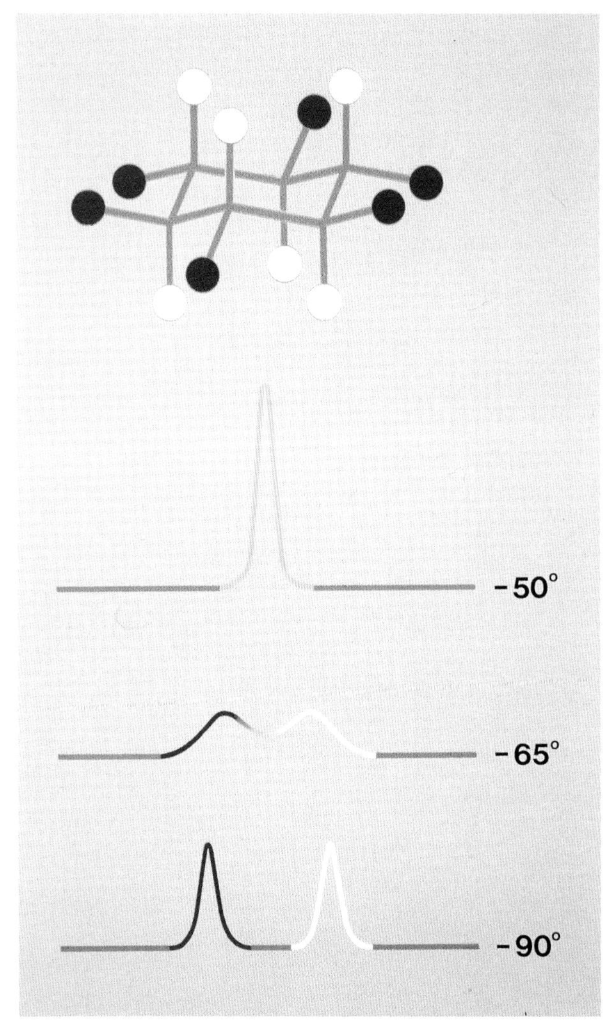

Nuclear magnetic resonance signals of equatorial (red) and axial (white) hydrogen nuclei of cyclohexane as a function of temperature.

intensity is proportional to the number of atoms in that environment. This is shown, somewhat simplified, for cyclohexane in the spectrum at the bottom of the adjacent Figure. The red outlined signal corresponds to the equatorial hydrogen atoms, the white outlined signal to the axial ones.

Nuclear resonance spectroscopy is a relatively 'slow' method compared with the frequency of the inversion process; in our example with the photographs of the rotating vase, it would correspond to a time-exposure. In order to make a clean 'photographic' separation of the axial and equatorial hydrogen atoms, the speed of the inversion process has to be slowed down. This was accomplished for the bottom spectrum in the adjacent Figure by cooling the sample to –90 °C.

When the temperature of the sample is raised, the inversion frequency of the cyclohexane molecule increases, and the spectrum, like a photograph of a fast moving object, becomes blurred. This is seen in the spectrum at –65 °C. Finally, at temperatures above –50 °C, the inversion frequency is so high that the molecule swings backwards and forwards through the inversion process innumerable times during the 'exposure time'. Our measurement method then produces a signal corresponding only to an averaged out molecule and hence it can no longer distinguish between the two types of hydrogen atoms. This is shown in the top spectrum of the previous Figure.

From the symmetry standpoint we thus find ourselves in a dilemma. At very low temperatures our measurement method tells us unequivocally that cyclohexane contains two kinds of hydrogen atoms, axial and the equatorial, indicating a lower symmetry with a C_3 rotation, as shown in the upper part of the previous Figure. At high temperatures, the same measurement method indicates a structure in which all twelve hydrogen atoms are equivalent, as would only be the case for a planar hexagonal molecule, as long as one thinks in terms of rigid, inflexible models. As we see, this conclusion would be incorrect, for we know that we are here dealing with an artefact caused by the slow method of measurement which can produce only an averaged, blurred picture of the molecules. At any instant the molecule has a lower symmetry than that indicated by the averaged picture.

This example makes clear that the manner in which one assesses the distinguishability or indistinguishability of the initial and final states after carrying out a rotation or reflection operation has a decisive influence on the answer. According to the method and experimental conditions, the answer could be 'yes' or 'no', and it is quite possible that both answers are meaningful. This will depend on what we want to know and on what characteristics of the molecule we focus our attention.

So far we have been tacitly assuming that molecules are rigid objects, like vases or double-bladed swords or lollipops or other objects whose symmetry properties we earlier brought into the discussion. This assumption is far from the truth and sometimes very far from the truth. All molecules have some degree of flexibility. At the very least, the atoms vibrate about their equilibrium positions, sometimes with considerable amplitudes, and in more flexible molecules it is sometimes questionable whether one can speak of a molecular 'structure' at all. As far as the more rigid molecules are concerned there is usually no problem in deciding what the molecular symmetry is. For a flexible molecule, however, there is the question of the temperature at which the observation of some molecular property is made and the 'shutter speed' of the observation concerned; for example, at low temperature cyclohexane appears to have two kinds of hydrogen atom, while at higher temperature only one kind as we observe only the 'average' of two rapidly interconverting structures. Or consider a molecule with a trigonal bipyramidal frame with five different types of atoms at the vertices. In the rigid arrangements shown in the upper Figure on page 91 the molecule has no symmetry, but if the molecule undergoes rapid interchange of equatorial and axial substituents, as described in Chapter VII, then, after some time, the averaged molecule would appear to have the full symmetry of a trigonal bipyramid but with one fifth of each type of atoms at each vertex – interesting as a mathematical concept but physically unreasonable.

Another complication in the description of the symmetry of non-rigid molecules can be illustrated by the toluene molecule **38**. In the conformation shown below, the molecule has mirror reflection as its only symmetry operation; all the atoms except two hydrogens of the methyl group lie in a common plane, and these two hydrogens are related by reflection across this plane; thus atoms b and c are symmetry equivalent but not atom a.

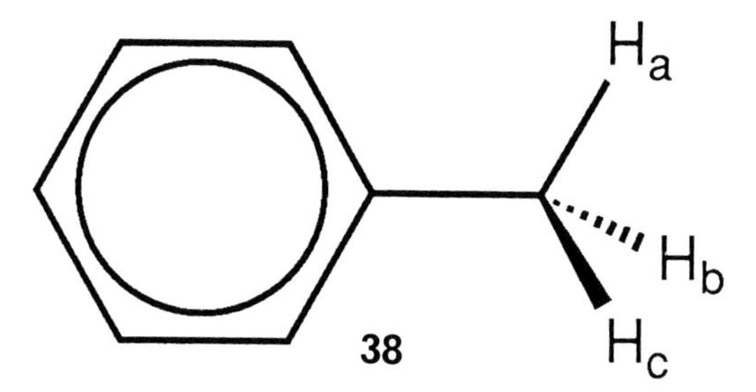

Now imagine that the methyl group can rotate by 120° jumps about the exocyclic C–C bond so that its three hydrogen atoms change places. Insofar as hydrogen atoms are indistinguishable, this operation converts the molecule into itself, or, in other words, the initial and final states of such a rotation cannot be distinguished. But this is just the characteristic that we used to define a symmetry operation in Chapter V. Nevertheless,

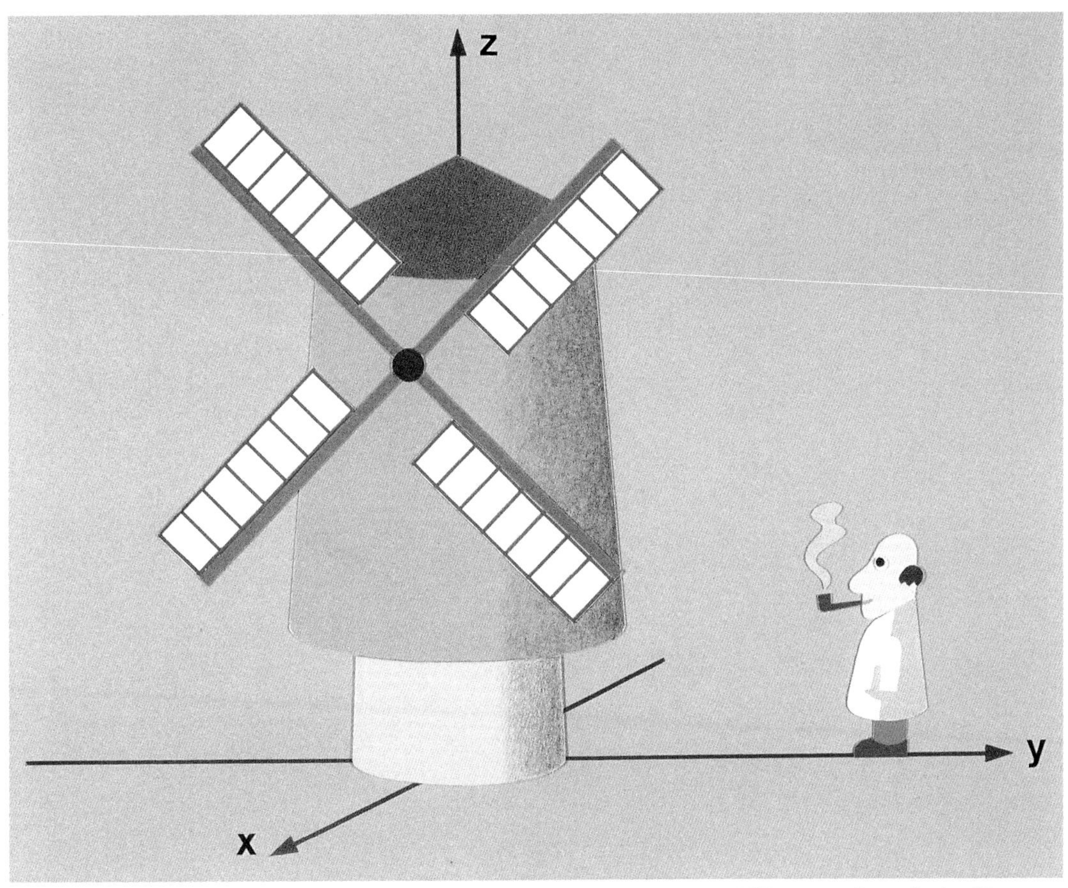

Windmill, an object showing symmetry due to internal rotation. Whereas the windmill as a whole belongs only to the point group C_1, the internal rotation of the wing by 90° – or multiples of 90° – transforms the windmill into itself.

we have to admit that this kind of internal rotation is intrinsically different to the rotations, reflections, and inversions that we have been considering up to now as examples of symmetry operations. These involved no change in the relative positions of the various parts of the object, whereas our new operations involve motion of one part of the object relative to the rest of it.

As we saw at the end of Chapter V, the usual symmetry operations such as rotations and reflections can be regarded as mere transformations of the coordinate system, and it is clear that the internal rotation of the methyl group of toluene cannot be described by such a transformation. A more detailed analysis of this kind of problem would take us into very deep waters, and many of the questions raised have not yet been answered to everyone's satisfaction.

IX.

Der Kristall ist ein chemischer Friedhof.

Leopold Ruzicka

Let us examine a crystal..... the equality of the sides pleases us; that of the angles doubles the pleasure.

Edgar Alan Poe

For a long time, to speak with the words of Leopold Ruzicka, crystals were thought of by chemists as nothing other than 'chemical cemeteries'. Under the influence of modern solid-state chemistry and physics, this attitude has changed dramatically, especially since it has become apparent that an exact knowledge of the reactivity and dynamical properties of solids is vital for modern technology; one need only think of the importance of plastic materials and of solid-state electronics. Whereas chemical reactions in solids were once regarded mainly as a nuisance, solid-state chemistry is today an acknowledged field, on which respectable international conferences with hundreds of participants are held.

However, Ruzicka's metaphor of the 'chemical cemetery' has a definite pictorial appeal: in a crystal the individual molecules are neatly lined up with their neighbors in tidy rows; although the molecules in a crystal are not motionless, they do not move far from their positions, quite different from the molecular mazurkas that can be imagined to take place during chemical reactions in solutions or in gases. The properties of a crystal depend not only on the structure of the molecules out of which it is built but also on the details of the way in which these molecules are arranged and even on the defects and disturbances that may occur in such arrangements. In this chapter we enquire how the relative positions of the individual molecules to one another can determine the

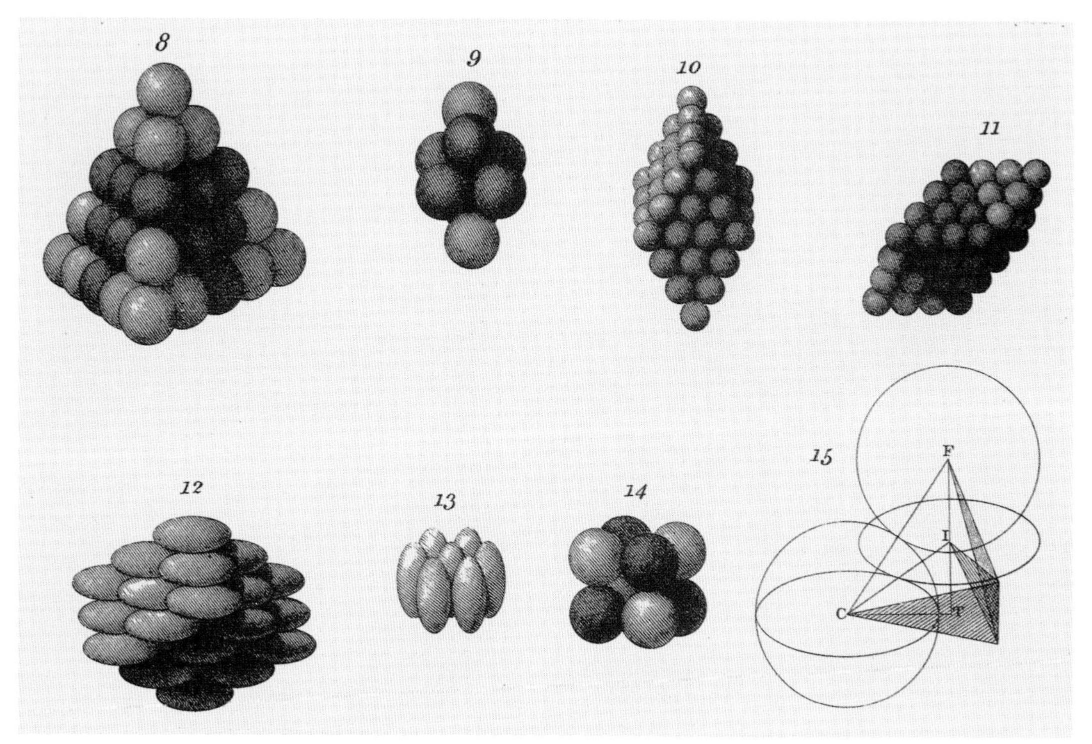

Various types of packing, from Wollaston's Bakerian lecture[11].

structure and symmetry of the crystal as a whole. One example has already been encountered: how Kepler's ideas about the closest packing of spheres led to valid models for the structure of many metals. If one abandons the idea that the elementary building blocks of the crystals must be spheres, as was done by William Hyde Wollaston in his 1812 Bakerian Lecture, from which the above picture is borrowed, then the diversity of possible macroscopic crystal symmetries is enormously increased.

The science of crystallography can be said to have begun with the observation of the Danish anatomist and geologist Niklaus Steno (1638–1686) that although quartz crystals may differ widely in appearance, the angles between corresponding faces are always the same from one specimen to another. Similar observations were made about the same time for Iceland Spar (calcite) by Steno's fellow countryman Erasmus Bartholin (1625–1698), the discoverer of double refraction produced by this mineral. These findings were corroborated by Jean Baptiste Rome de Lisle (1736–1770), who noted that the interfacial angles differ from substance to substance but are characteristic for a given substance. Thus, by the middle of the 18th century it had been recognized that although natural crystals of a given substance can occur in very different shapes, they show the same pattern of interfacial angles. In 1784 Abbé René Just Haüy (1743–1822) published his book *Essai d'une théorie sur la structure des cristaux appliqué à plusieurs genres de substances cristallines*, in which he introduced the concept of space lattices and showed that this constancy of angles could be explained if the naturally occurring bounding planes of the crystal were assumed to be those that contain many lattice points. A similar phenomenon can be observed if one walks through a vineyard where the vines are planted in a regular array; certain directions stand out as containing prominent rows. Haüy suggested that crystals were built by regular repetition of fundamental units of pattern, but he did not believe that the ultimate particles were necessarily spheres; for him they could be of any arbitrary shape, and he called them 'molécules intégrantes'. The fact that crystals are symmetric and not just any old shape puts restrictions on the possible arrangements. This leads to the purely geometric problem of finding all the types of symmetry that are possible by arranging a collection of identical objects in space.

By the end of the 19th century, the geometrical theory of space lattices had been completed, leading to the recognition that there are only a finite number of ways of combining symmetry operations with translations to make periodic patterns. In addition to the symmetry operations we have met up till now, new combinations are possible; screw axes, combinations of rotation with translation parallel to the rotation axis, and glide planes, combinations of reflection with translation parallel to the reflection plane. There are, in fact, exactly 230 such combinations, known as space groups, as was found almost simultaneously by Evgraf Stepanovich Fedorov (1853–1919) in Moscow, and Artur Moritz Schoenflies (1853–1928) in Göttingen, working quite independently of each another and using quite different approaches to the problem. The mathematical theory of crystallography had been completed, but, at the same time, nothing definite was known about the shapes of the particles on which the symmetry operations were sup-

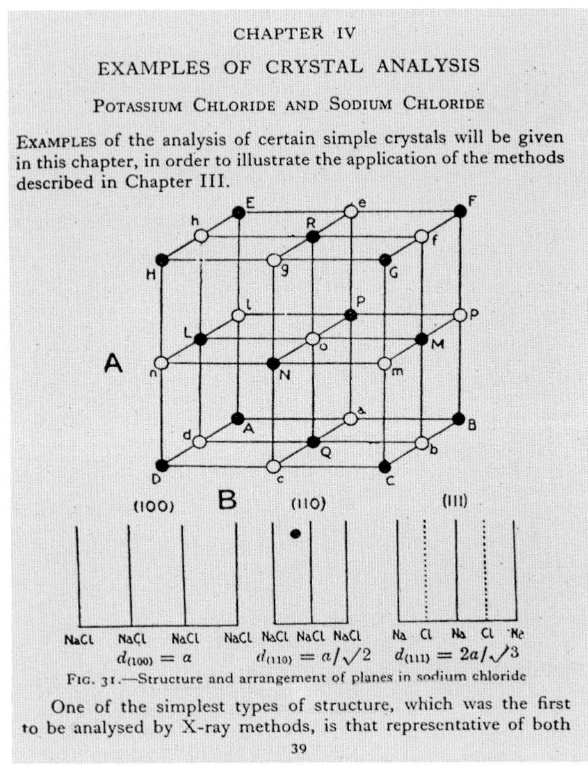

CHAPTER IV

EXAMPLES OF CRYSTAL ANALYSIS

POTASSIUM CHLORIDE AND SODIUM CHLORIDE

EXAMPLES of the analysis of certain simple crystals will be given in this chapter, in order to illustrate the application of the methods described in Chapter III.

$d_{(100)} = a$ $d_{(110)} = a/\sqrt{2}$ $d_{(111)} = 2a/\sqrt{3}$

FIG. 31.—Structure and arrangement of planes in sodium chloride

One of the simplest types of structure, which was the first to be analysed by X-ray methods, is that representative of both

39

Crystal structure of sodium chloride, from W. L. Bragg's The Crystalline State, *1937.*

posed to operate, i. e., the 'molécules intégrantes'. It was only after the discovery of X-ray diffraction in 1912 by Max Theodor Felix von Laue (1879–1960) that it became possible to look into the inner structure of crystals, as was achieved within a few months by William Henry Bragg (1862–1942) and his son William Lawrence Bragg (1890–1971) for diamond and simple ionic solids.

Among the very first crystals to be examined by the new method were those of the alkali halides, sodium chloride, for example, which was shown to be built from an alternating pattern of spherical sodium and chloride ions, just like a chess board – but in three dimensions.

There were no discrete NaCl 'molecules' to be seen. It may be hard to imagine today the extent to which traditional chemical concepts were upset by these results. In particular, the English chemist Henry Edward Armstrong (1848–1937), who was noted for his strong and outspoken views, made no secret of his unqualified rejection. In a letter to the English journal *Nature* he wrote[35]:

> "Prof. W. L. Bragg asserts that in sodium chloride there appear to be no molecules represented by NaCl. The equality in numbers of sodium and chlorine atoms is arrived at by a chess-board pattern of these atoms; it is a result of geometry and not of a pairing-off of these atoms.....Chemistry is neither chess nor geometry, whatever X-ray physics may be.....It were time the chemists took charge of chemistry once more and protected neophytes against the worship of false gods; at least taught them to ask for something more than chess-board evidence."

Later, of course, it became clear that such highly symmetric structures are adopted only by the simplest ionic compounds and metals. Crystals of most compounds generally have much lower symmetry and have structures in which the individual molecules are clearly recognizable as discrete entities. Chemists have learned to live with the results of crystal structure analysis, and, indeed, to depend on them. But even although the detailed atomic arrangements in more than 100,000 crystals are known today, the general question, how does the shape of individual molecules determine the structure and symmetry of the crystal, is still unanswered. Given the molecular structure, it is still not possible to predict the way in which the molecules will arrange themselves in the crystal. This is a problem of physics, one where the difficulty lies in its sheer complexity rather than in any conceptual obstacles. The purely mathematical problem, what are the possible symmetries that can be produced by periodic repetition of patterns, has been solved and is an ideal training ground for anybody with an interest in the systematic application of symmetry arguments. Here, some simple preliminaries may give an idea of what is involved.

We turn first to the simplest case, the generation of a line pattern, which we already met in Chapter II. We start with a completely unsymmetrical shape, for example, a gnome, of the type known in German as a Gartenzwerg and portrayed in the left half of the Figure below. Reflection of such a 'left' gnome necessarily produces a 'right' gnome; these two non-superimposable gnomes will now serve us as a basis for the generation of all possible line patterns.

Relative to our previous symmetry arguments about individual objects or molecules, the new aspect is that the patterns of interest are now to be generated by periodic repetition of the same unit, our gnome, for example, at equal distances along a line. The top strip of the next diagram shows this for the most trivial case, that of simple repetition.

Left and right gnomes.

The choice of additional symmetry operations is, of course, rather limited in a line pattern. We can rotate our gnome by 180° about an axis perpendicular to the translation direction (C_2, second row of the diagram) or we can reflect him across a line which must be either perpendicular to the translation direction (third row) or coincides with it (fourth row). Finally, we can try to combine these additional symmetry operations. By trial, or by appropriate mathematical methods, one can show that there are, in principle, only seven different types of such one-dimensional strips, all of which are shown in the accompanying Figure. This means that although the motif, here our gnome, can be anything imaginable, there are only seven ways, so-called line groups, in which it can be repeated to make a symmetrical periodic linear pattern.

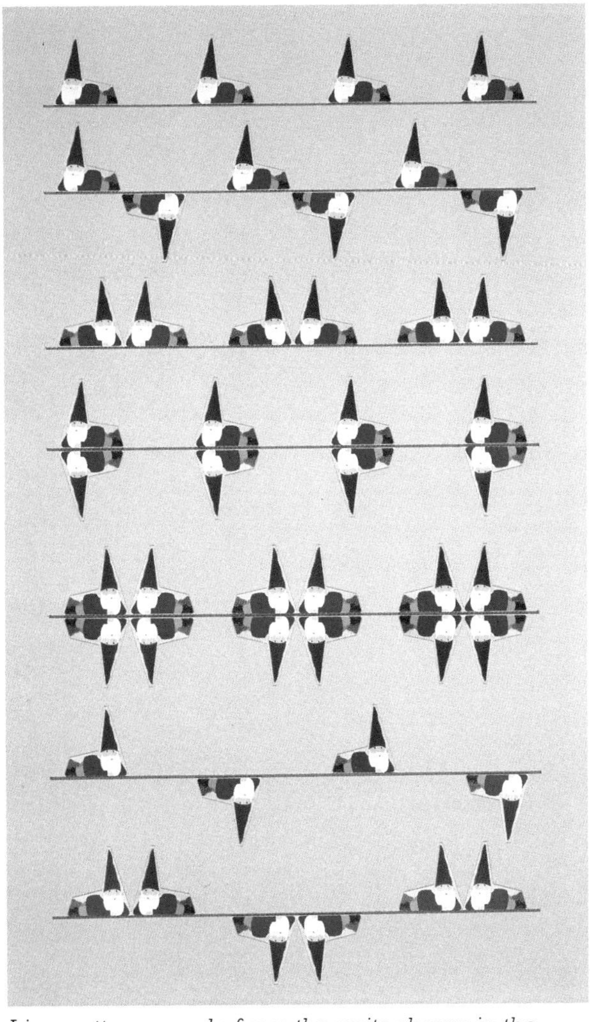

Line patterns made from the units shown in the previous Figure.

A two-dimensional periodic pattern occurs when a given motif of pattern is repeated periodically along two nonparallel directions, as is familiar to us from wallpapers and Christmas wrapping papers. We choose some arbitrary point in one motif and consider the arrangement of all equivalent points obtainable by pure translation of the original point. This arrangement constitutes a net that has the same periodicity as the pattern itself. Or we can think of the net as a device that converts a single motif into the entire pattern. The position of the net can be chosen arbitrarily; it is only the relative arrangement of the net points, the lattice points, that matters. The net can be characterized, as shown in the adjacent Figure, by a parallelogram whose sides may be chosen along two translation directions – a unit cell. There is no unique way of choosing such a unit cell, but, regardless of this arbitrariness, the unit cell abstracted from the periodic pattern in this way and defined by the lengths of its two sides and the included angle is already a first, important classification characteristic of such a two-dimensional periodic pattern. Three examples are shown in the upper part of the next Figure – the general case of a lattice defined by an arbitrary parallelogram, and two special cases in which the unit cell is rectangular (angle of 90° between the edges) or square (equal edges and 90° angle).

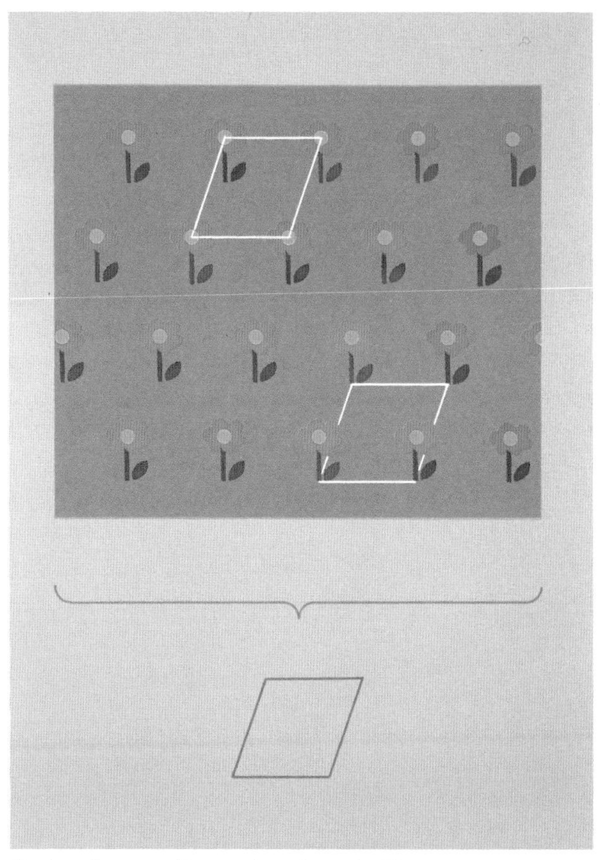

A simple two-dimensional periodic pattern; the position of the unit cell is arbitrary.

As might have been expected, the number of possible ways in which additional symmetry operations can be combined with the translation operations is larger than in the one-dimensional case, but is not unlimited. Indeed, it can be shown without too much difficulty that, in spite of the apparently infinite variety of possible wallpaper patterns, there is only a finite number of essentially different types, known as plane groups – 17 to be precise, a number that may strike one as being astonishingly small! One reason for this drastic limitation is that the possible rotations in such a two-dimensional pattern are restricted to C_2, C_3, C_4, and C_6.

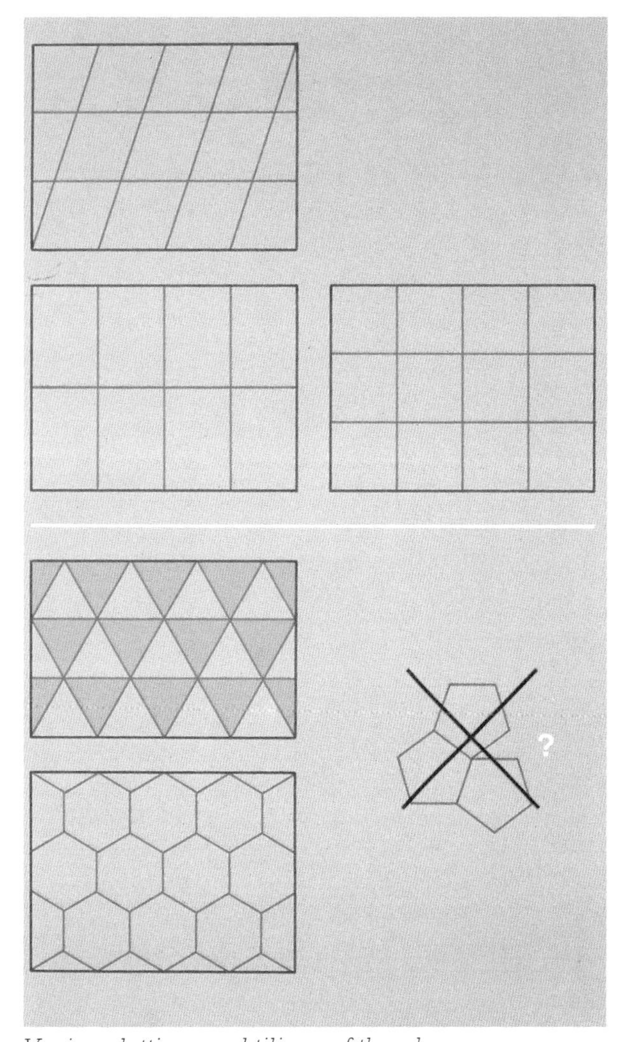

Various lattices and tilings of the plane.

In particular, C_5 is forbidden, as one can easily convince oneself from the impossibility of covering a bathroom wall with regular pentagonal tiles (shown in the adjacent Figure, lower right). It is possible only with equilateral triangles (C_3), squares (C_4), or hexagons (C_6). The three-dimensional case can be analyzed in an analogous way, leading to the 230 essentially different combinations of symmetry operations that are possible in regularly repeating patterns – so we have 7 line groups, 17 plane groups, and 230 three-dimensional space groups.

A related but somewhat simpler problem is to enumerate the possible external symmetries of crystals, the crystal classes, or crystallographic point groups: the combinations of symmetry operations, excluding translations, that are consistent with the allowed rotation axes. The mathematical proof that there are only 32 such crystal classes was given by Johann Friedrich Christian Hessel in 1830. The following illustration is taken from his major work *Kristallometrie oder Kristallonomie und Kristallogaphie*, published in Leipzig in 1831 but largely ignored and then forgotten until the end of the century.

Crystals can be assigned to the correct class on the basis of their external shape, but it was only after the development of X-ray analysis, when information about the detailed internal structure became available, that crystals could be assigned to the correct space group. As an example, we show the image of a projection through a crystal of hexamethylbenzene (Figure on the next page), first solved by Kathleen Lonsdale (1903–1971), from which not only the structure of the individual molecules, corresponding to the formula **39**, is apparent but also the manner in which these molecules are arranged in the crystal lattice.

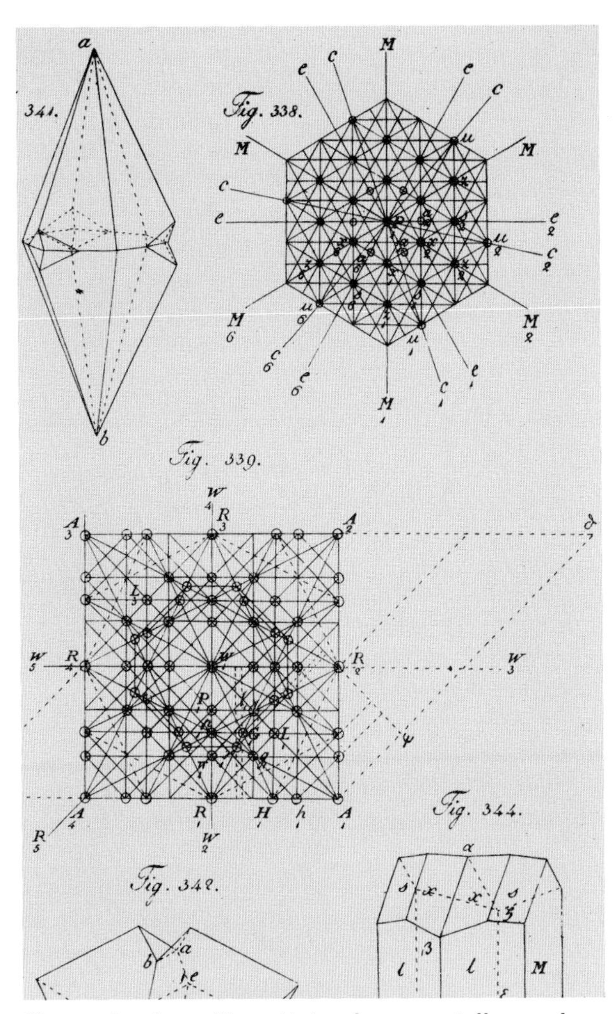

Illustration from Hessel's book on crystallography.

Naturally, if more complete information about the detailed structure of the crystal were required, this image would have to be complemented by a projection along some other direction or by various sections through the structure. From such a series of sections, a three-dimensional model of the crystal structure can be built. Such models used to be constructed from balls, representing the atoms, and wooden or metal rods, representing the bonds, but these have now been superseded by computer graphics.

CH_3 CH_3 CH_3 CH_3 CH_3 CH_3

39

The determination of the structures of many thousands of crystalline compounds

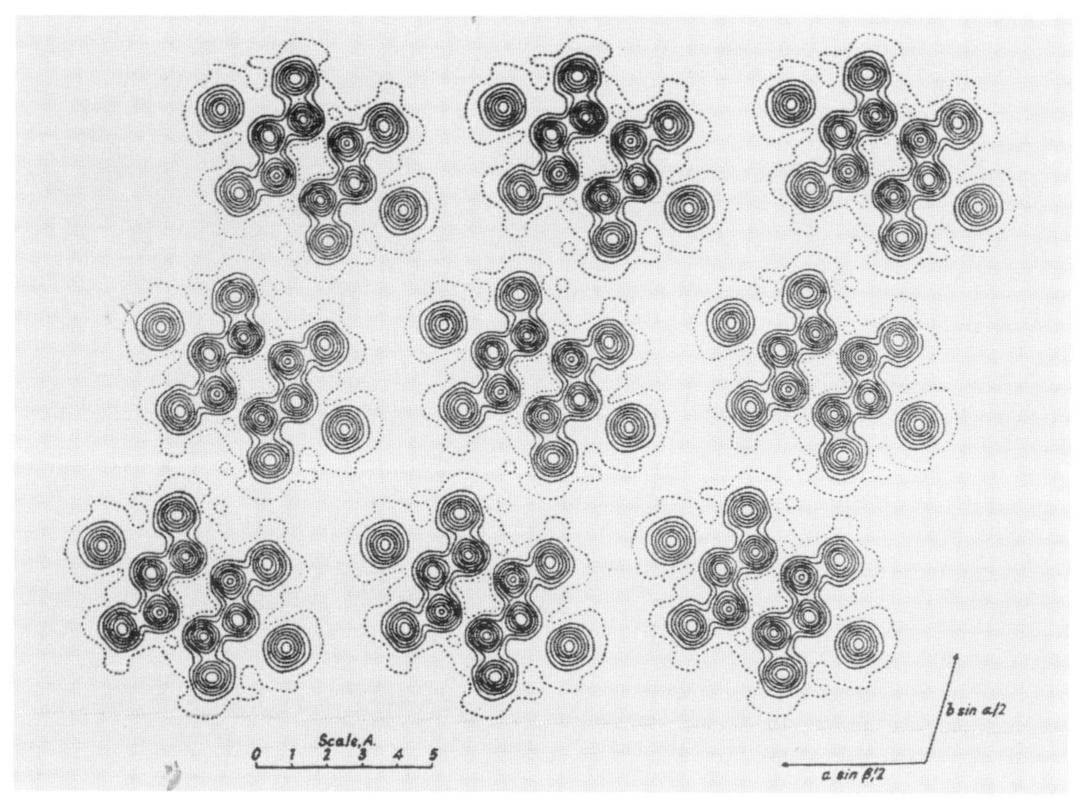

Projection of the hexamethylbenzene crystal structure[36].

showed that the distribution of crystals among the 230 space groups is far from uniform. Among molecular crystals, the half-dozen most common space groups account for about 90% of known crystal structures. Some symmetry elements (such as inversion centers) tend to produce efficient packings of molecules, while others (such as sixfold rotation axes) do not. Of course, compounds that consist exclusively of molecules of one sense of chirality can crystallize only in space groups that contain proper symmetry operations. There are 65 of these chiral space groups, leaving 165 racemic space groups that contain improper symmetry operations as well as proper ones. A racemic solution, containing equal numbers of molecules of opposite senses of chirality, can, in principle, crystallize either as a racemate in one of the 165 racemic space groups or as a conglomerate, a mixture of enantiomerically pure crystals of opposite sense of chirality. The additional packing possibilities among the racemic space groups mean that the former possibility is more likely to occur. Crystallization from a racemic solution usually leads to racemic crystals, but there are many exceptions.

With the enumeration and description of the 230 space groups it appeared that the rules of formal crystallography had been established for all time. It came as a surprise – or even as a shock – when, in 1984, the Israeli physicist Dany Shechtman[38] produced an aluminium-manganese alloy that *seemed* to be a crystalline compound with fivefold symmetry. At least, the diffraction pattern of the alloy indicated a three-dimensionally periodic structure with fivefold symmetry, contrary to all the rules of classical crystallography. Other examples of apparently crystalline materials with 'forbidden' symmetries were soon found. It is now recognized that although such materials cannot be periodic and perfectly symmetric they come close to it; they are known as quasi-crystals.

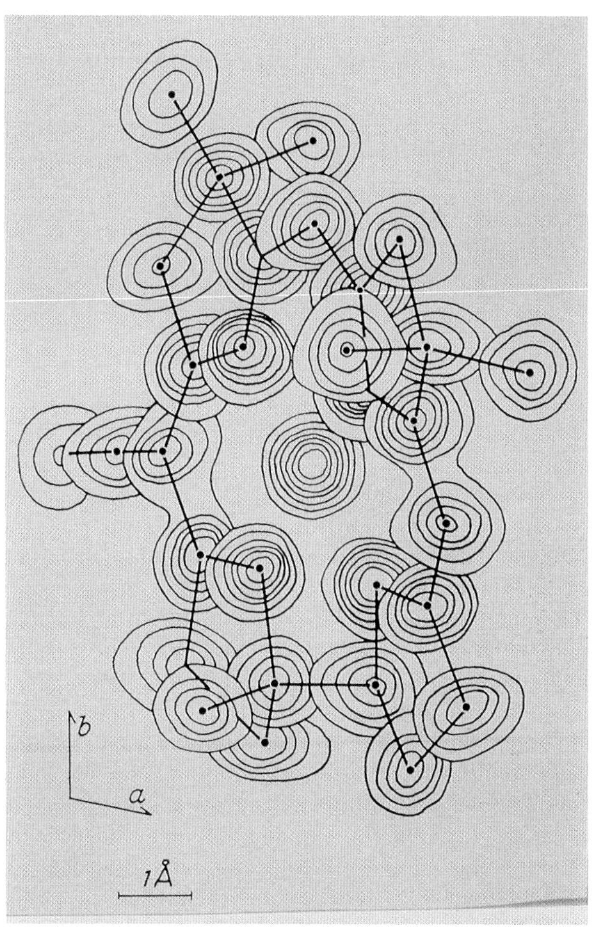

Electron density map of the molecule of the first synthetic compound containing the vitamin B_{12} nucleus, prepared at the ETH-Zurich in the laboratory of Albert Eschenmoser[37].

The discovery of quasi-crystals has led to the recognition of quasi-periodic structures, the existence of patterns that never repeat themselves but are not random. Such patterns are to be found in certain tilings of the plane. As we have already seen in our discussion of two-dimensional periodic patterns, one can completely cover a plane with identical tiles shaped as parallelograms, equilateral triangles, squares, or regular hexagons – but not as regular pentagons, since a

Electron diffraction pattern showing tenfold symmetry.

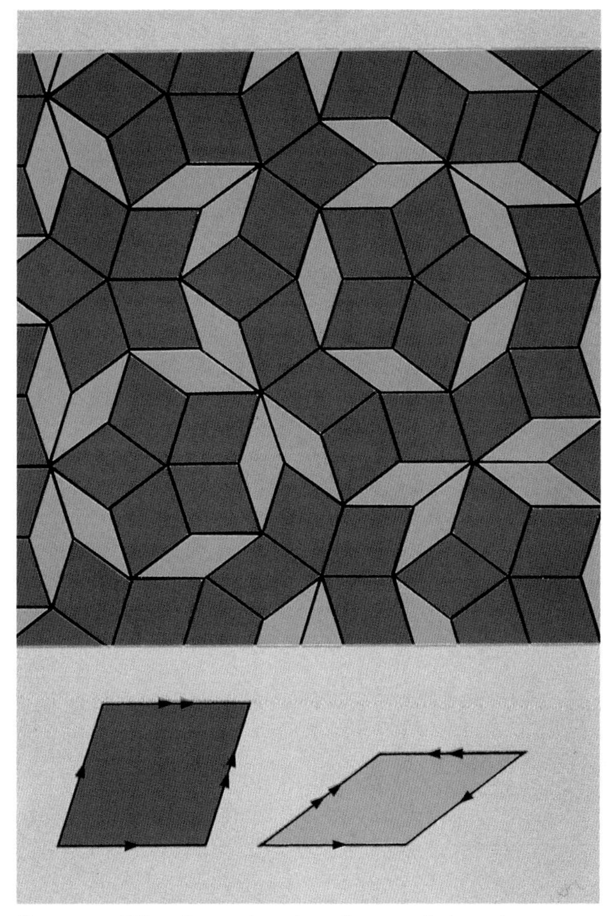

Penrose tiling based on two types of rhomb.

planar net with fivefold symmetry is impossible. A few years before the discovery of quasi-crystals, the English mathematician Roger Penrose showed that if two shapes of tiles are allowed, tilings can be made that are almost periodic and show almost fivefold symmetry. An example is shown in the accompanying Figure. It is based on two kinds of rhombs with equal sides, one with angles of 72° and 108°, the other with angles of 36° and 144°, which have to be fitted together according to a 'matching' rule; the single and double arrows drawn on the edges of the rhombs in the lower part of the Figure must be of the same kind (both single or both double) and in the same direction.

When one tries to make an actual tiling with these two shapes of tile, one soon finds that this apparently simple child's game has fantastic levels of complexity. Try it! Make some outlines of the two shapes of tile and try to fit them together according to the rule mentioned above. You will begin by successfully tiling a small area and then possibly find that you have to retreat a little in order to make further progress. The arrangement at any given point depends on the arrangement at more distant parts of the pattern, so you will need to think ahead. You will also find out that there is not a unique way of arranging the two kinds of tiles, there are an enormous number of different arrangements even for quite small areas to be covered. In fact, the number of ways of tiling the whole (infinite) plane is of the order of the infinity of real numbers, that is, infinitely more than the infinity of integers. The pattern as a whole has no symmetry whatsoever. It never repeats but it is obviously far from random; there are portions that have local periodicity and local fivefold symmetry. The diffraction pattern of a Penrose tiling has sharp reflections characteristic of a periodic crystal and has fivefold symmetry. In three-dimensions, two shapes of rhombohedra can be packed together to fill space in an analogous way to that in which the tiles can cover the plane.

Although the calculated diffraction patterns of quasi-periodic structures represented by Penrose tilings are similar to those produced by quasi-crystals, it is still not at all clear whether such tilings have any actual physical relevance. There are still many unsolved and controversial problems concerned with the nature of quasi-crystals. They are probably not, as had been once proposed, a 'new form of matter' but they have opened our eyes to the rich mathematical and physical possibilities offered by quasi-periodic structures.

Many properties of a crystal, especially its electric properties, are largely determined by its symmetry, which is itself a consequence (not yet fully understood) of the forces holding the atoms or molecules together in the particular arrangement they adopt in the lattice. If the temperature of a crystal is gradually raised, there comes a moment when the thermal motion of the individual particles becomes so great that it overcomes the attractive, ordering forces; the crystal begins to melt into a liquid. In this process, the beautifully regular order, and the symmetry associated with it, is lost, as indicated schematically in the next Figure for a two-dimensional crystal with dense packing of spheres.

Although the individual particles in the liquid are still largely in direct contact with each other, and the immediate surroundings of any given molecule still possess at least a certain approximate symmetry, the destruction of the overall symmetry means that those properties of the crystal that depend on its symmetry are lost. In practice, it is this loss of symmetry that makes the study of liquids one of the most difficult chapters of physics and chemistry.

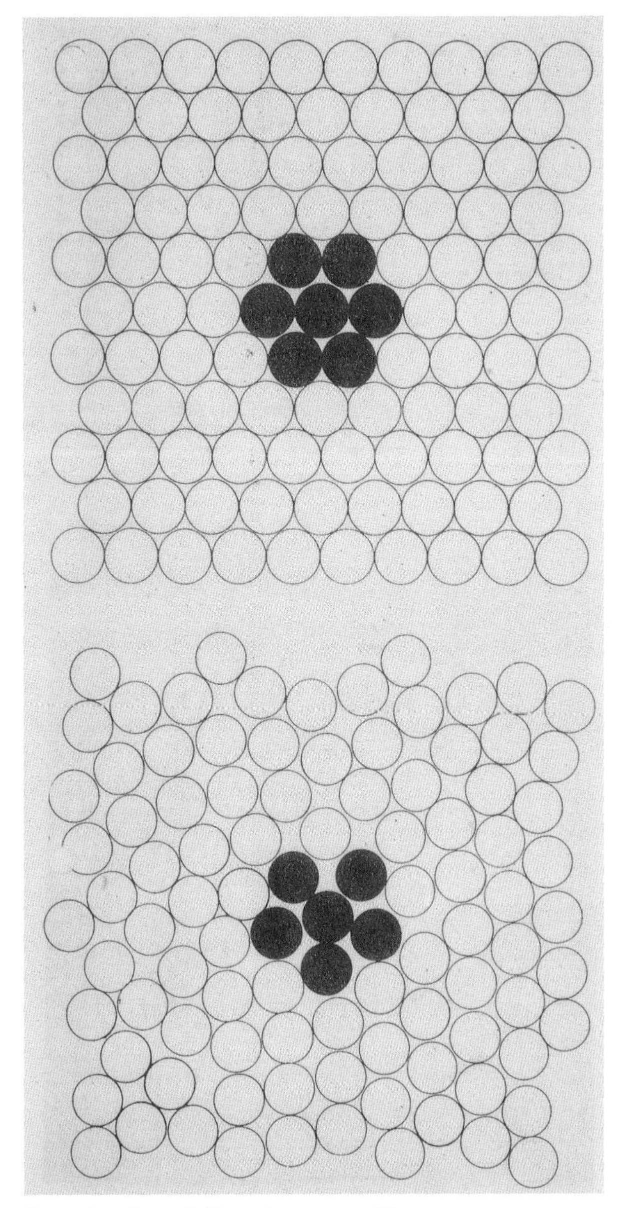

Introduction of disorder on melting.

The fact that symmetry is produced by the regular arrangement of some motif and destroyed when this regularity is disturbed may suggest that symmetry must be involved somehow in problems concerning the relationship between order and disorder.

Before we examine this question, we must go backwards a little and attempt to define more exactly what we understand by 'disorder'. A child's playroom is a good place to begin. Let us assume that six toys have to be accommodated in a chest containing six drawers. Maximum order is achieved when all six toys are located in one of the bottom boxes, as indicated in drawing *A* of the next Figure.

If we wish to play with the ball, for example, we need only open this particular box and we have it immediately. A first step towards 'disorder' would be made if we put one of the toys – we don't specify which – in the second box as shown in drawing *B*. As we have six different toys, there are six different ways to do this. If we distribute the toys over three or more boxes, the degree of 'disorder' increases, which means that if we are ignorant of the particular distribution we may have to open several boxes in order to find any given toy. Maximum 'disorder' is achieved when there is one toy in each box; there are $6 \times 5 \times 4 \times 3 \times 2 = 720$ ways of achieving this, one of which is

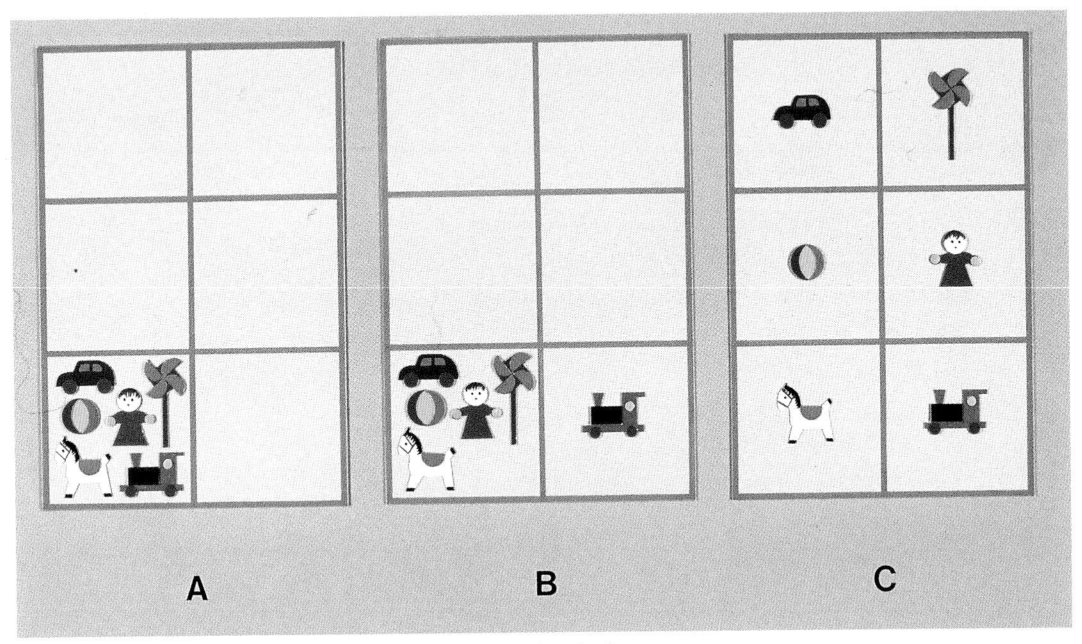

Gradual introduction of disorder of six toys in six boxes.

shown in the drawing *C*. We take the number of possible arrangements, 1 for situation *A*, 6 for *B*, 720 for *C*, as a measure of the degree of 'disorder' in this simple system, consisting of six boxes and six toys, and we designate this number by the Greek letter Ω. This approach, introduced by the Austrian physicist Ludwig Eduard Boltzmann (1844–1906), has proved to be remarkably fruitful in dealing with the problem we are discussing. With increasing number of toys and drawers, the number of possible distributions Ω increases rapidly; applied to chemistry, with its unimaginably large numbers of molecules and of states in which they may occur, the number of distributions becomes so enormous that we use the logarithm $\ln\Omega$ instead of Ω itself as a measure of disorder. In molecular systems disorder increases with temperature, and, to make the connection with measurable quantities, $\ln\Omega$ is multiplied by a constant k to give a quantity $S = k\ln\Omega$, known as the *entropy* of the system. The formula $S = k\ln\Omega$, one of the most important in the whole of physics, is inscribed on Boltzmann's gravestone in Vienna.

Away from the playroom, a rough analogy to the above example can be found in the behavior of the physical states of matter. A crystal containing a large number of molecules can be taken – *cum grano salis* – to correspond roughly to the situation shown in the next Figure. When the crystal melts,

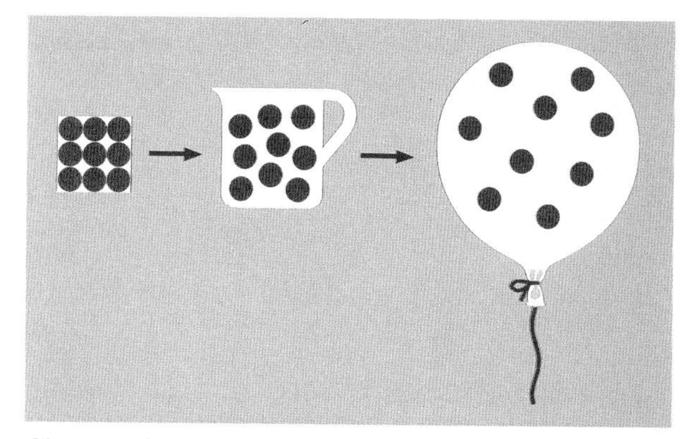

Changes in state: crystal, liquid, gas.

the molecules usually (not always – remember that ice floats on water) require a little more room, and finally, when the liquid evaporates to a gas, we have again a system of maximum disorder. The changes in entropy for the transformations crystal→liquid and liquid→gas can be measured; in the first step the entropy increases by about 25 to 40 units (we ignore the question of how such units are to be defined) and in the second step it is, as might be expected, much larger, about 90 units.

If we now believe that through this reference to our everyday experience we have the entropy concept under control, we will probably be disappointed by the next example. It seems reasonable to assume that the 'disorder' in a balloon containing helium and in one containing xenon is the same, under conditions of equal volume, pressure, and temperature and hence equal numbers of atoms in the two balloons. It turns out, however, that the entropy of the xenon balloon is about 35% larger than that of the helium balloon. This may suggest that other factors, such as the weight of the atoms, might play a role. Indeed they do. "The Lord is subtle", as Einstein said "but not malicious." One of the factors that influence the entropy is the molecular symmetry.

Substances containing highly symmetric molecules have a lower entropy, i.e., they are more 'orderly' than those containing low-symmetry molecules. In this connection, however, we need to express the amount of symmetry by a number. As mentioned earlier, this could be the number of distinct symmetry operations possessed by the molecule, but in the present context it is more relevant to take as the symmetry number Σ the num-

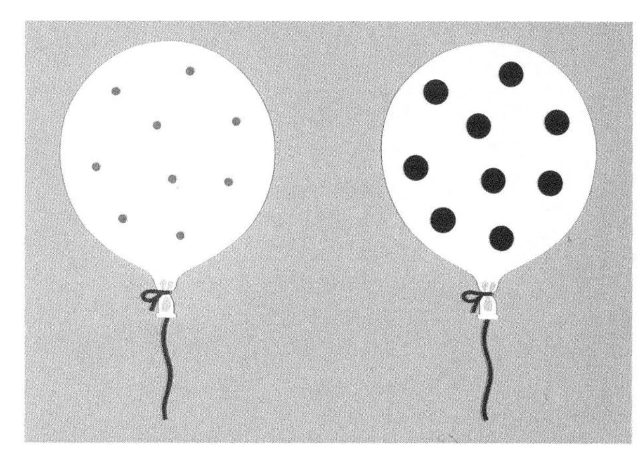

Equal volumes of helium and xenon at the same temperature and pressure do not have the same entropy.

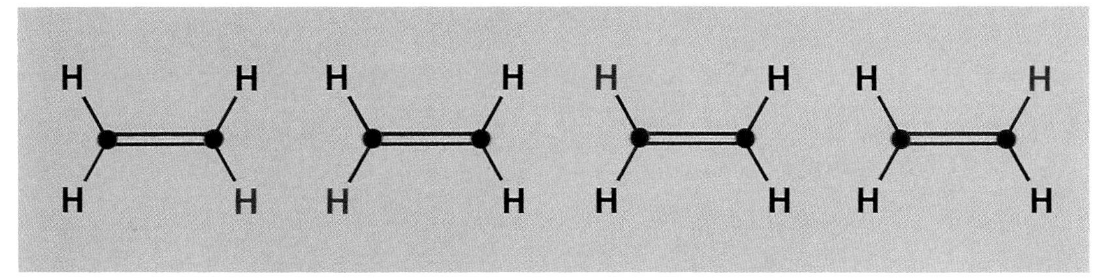

The symmetry number of ethylene is 4.

ber of indistinguishable ways the molecule can be oriented in space. As an example, we consider the ethylene molecule $CH_2=CH_2$. For counting purposes, we imagine that we can label one of the hydrogen atoms by coloring it red. One then sees that there are just four ways of placing the red atom, corresponding to four indistinguishable ways of orienting the molecule if we now imagine the red color to disappear. The symmetry number Σ thus equals 4 (Figure above).

Symmetry numbers for a few other molecules are listed below.

Molecule	Σ
CO	1
N_2	2
Ethylene	4
Methane	12
Benzene	12
Chlorobenzene	2
1,2-Dichlorobenzene	2
1,4-Dichlorobenzene	4
Cubane	24

We see that we get the right symmetry number Σ if we count only the rotation operations and forget about the reflections. Remember that, unlike rotations, reflection operations are not operations that one can actually carry out on objects, and the same applies to inversion of an object through a point in space. For these operations, one needs either mirrors or mathematics. This is why they are called improper symmetry operations, in contrast to rotations, which are proper symmetry operations.

For two isomers in chemical equilibrium, the one consisting of molecules with lower symmetry will be present in excess – other things being equal, as they are so often assumed to be in arguments of this kind, and as they so rarely are. Indeed, the amounts of the two isomers will be inversely proportional to their symmetry numbers Σ. We discuss an idealized example, that of the isomeric trichlorobenzenes, whose symmetry numbers are: 1,2,3-trichlorobenzene (**40**), $\Sigma = 2$; 1,2,4-trichlorobenzene (**41**), $\Sigma = 1$; 1,3,5-trichlorobenzene (**42**), $\Sigma = 6$. We may assume that the interconversion process consists of an exchange of a chlorine and a hydrogen atom within a given molecule, i.e. a shift of a chlorine atom from one position to another, as symbolically indicated by the arrows in the following schemes.

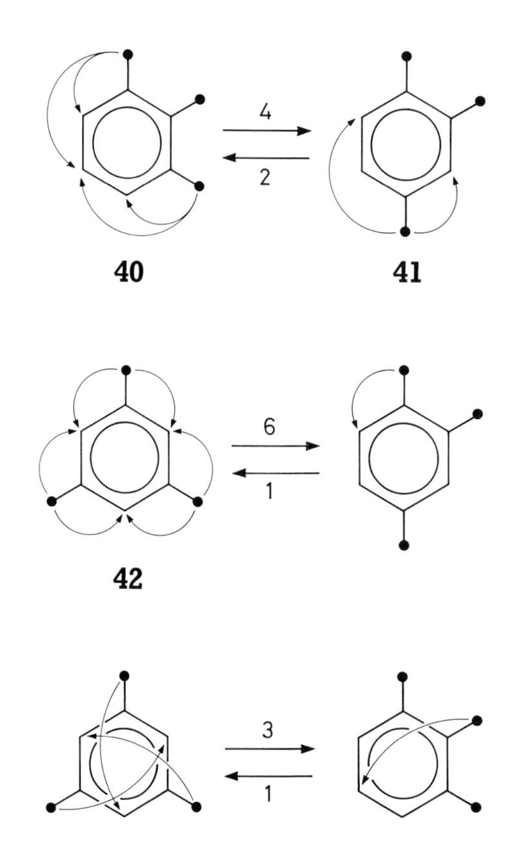

As can be seen, there are four shifts which yield 1,2,4 from 1,2,3, but only two which yield 1,2,3 from 1,2,4. Analogously, the number of ways in which the other transformations can be performed is: 6 for 1,3,5→1,2,4; 1 for 1,3,5→1,2,4; 3 for 1,3,5→1,2,3; 1 for 1,2,3→1,3,5. The ratios of the number of shifts for the forward and backward reactions are equal to the ratios of the symmetry numbers of the molecules involved, i.e. $4/2 = \Sigma(1,2,3)/\Sigma(1,2,4) = 2/1 = 2$, and so on.

If all the shifts are equally probable, then, at equilibrium the amount of 1,2,3, 1,2,4, and 1,3,5 will be in the ratios 3 : 6 : 1, i.e. inversely proportional to their symmetry numbers. Low symmetry isomers are thus preferred over high symmetry ones. These conclusions can also be reached by arguments depending on the entropy.

Indeed, with increasing symmetry number Σ the contribution to the entropy, that is, the 'disorder', decreases, and by quite a substantial amount. Thus, the entropy difference between the two isomers, *n*-pentane, $CH_3CH_2CH_2CH_2CH_3$, and neopentane, $C(CH_3)_4$, arising from the difference in the molecular symmetries, amounts to about 15 entropy units, a considerable fraction of the entropy change associated with the melting process.

The difference in melting point between these two isomers is striking. The more symmetrical one, neopentane, melts at –16 °C an abnormally high temperature for a hydrocarbon of this size, while the less symmetrical one, *n*-pentane, melts at –130 °C. The explanation of the abnormally high melting point of neopentane also has to do with symmetry. As we saw earlier, a liquid is more disordered and hence has a higher entropy than a crystalline solid. The neopentane molecule has a nearly spherical shape and can rotate in the crystal without destroying the crystalline order. In other words, some of the excess disorder characteristic of a normal liquid is already present in the solid. Hence the change in entropy on melting is smaller than usual. From thermodynamics we know that for two phases in equilibrium

$$\Delta G = \Delta H - T\Delta S = 0, \quad \text{or}$$

$$T = \Delta H / \Delta S.$$

The smaller the change in entropy, the higher the melting point – assuming that the heat of fusion is roughly the same in the two cases. Crystals that show this abnormal behavior on melting are called plastic crystals; their mechanical cohesion is weak and they can be regarded as a kind of intermediate stage between a crystalline solid and a normal liquid.

In a plastic crystal, orientational order is lost but translational order is largely maintained. Crystals containing long rod-shaped or plate-like molecules often show another kind of abnormal melting, corresponding to another kind of intermediate stage between crystalline order and liquid disorder. When these materials appear to melt they lose their translational order but retain their orientational order. The 'liquids' so produced flow but they have strongly directional properties. They are known as liquid crystals and have important industrial uses.

It is an inevitable consequence of the second Law of Thermodynamics that the entropy of the universe increases towards a state of maximum entropy. From this point of view it could appear that the Gods, unlike humans, have a long-term preference for low symmetry.

X.

Et sic angelus uno instante potest esse in uno loco et in alio instante in alio loco nullo tempore intermedio existente.

St. Thomas Aquinas

The central element of a conversation about symmetry in chemistry should really be concerned with the remarkable role that symmetry plays in the description of the electronic structure of molecules or, more exactly, of the motion of electrons. Unfortunately, even an elementary account comprehensible to the layman is impossible in such a brief space, mainly because the behavior of electrons, which we would like to imagine as little balls, scarcely allows any analogy to the behavior of those macroscopic objects that we encounter in everyday life. The 'common sense' that we have inherited through evolution and learned through daily experience is a block to any intuitive approach to the behavior of microphysical objects.

We all learn in school that light consists of electromagnetic waves of different frequencies, that is, different numbers of vibrations per second, and that a definite frequency corresponds to a definite color. The velocity of light waves is greatest in a vacuum where it has a value represented by $c = 3 \times 10^8$ meters per second. The velocity in dense media, such as glass or water is represented by u, which is smaller than c, while the frequency ν stays the same as in vacuum. The ratio of the two velocities $n = c/u$ is known as the refractive index of the medium. The following diagram illustrates how two light waves of different velocities but the same frequency propagate. The distance covered in a given time interval is proportional to the velocity so, for constant frequency, the wavelength must be inversely proportional to the index of refraction, that is, it is shorter in a dense medium than in a vacuum.

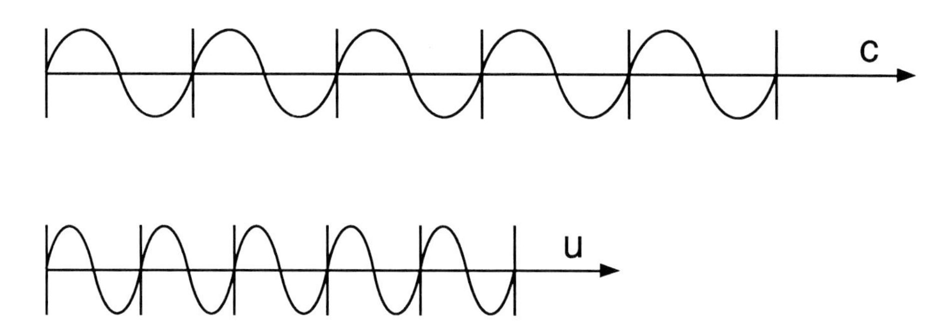

Since the light covers one wavelength λ for each completed vibration, it covers N wavelengths for N vibrations. The frequency ν is the number of vibrations per second, so we obtain for the velocities:

$$c = \lambda_o \nu, \quad u = \lambda \nu$$

We also learn in school that many problems in optics can be solved by a purely geometric analysis of the behavior of light rays, involving the tacit assumption that light can be regarded as a stream of particles, as suggested earlier by the great Isaac Newton (1643–1727). Is there a bridge between these two contradictory viewpoints? The answer is in the affirmative. It depends on a principle formulated by the French mathematician Pierre de Fermat (1601–1665), according to which a ray of light between two points always travels along the path that requires the shortest possible time. According to this principle, in a medium of variable refractive index, light will not go in a straight line but will minimize the time taken (not the path length) by bending into the region of lower refractive index where it will travel faster. The time Δt needed to traverse a path element of fixed length Δs is $\Delta s/u$ where u may vary from point to point. Fermat's principle now requires that the path taken from light source A to observation point B (see the next Figure) is the one for which

$$\Sigma(\Delta t) = \Sigma(\Delta s/u) = (1/\nu)\Sigma(\Delta s/\lambda)$$

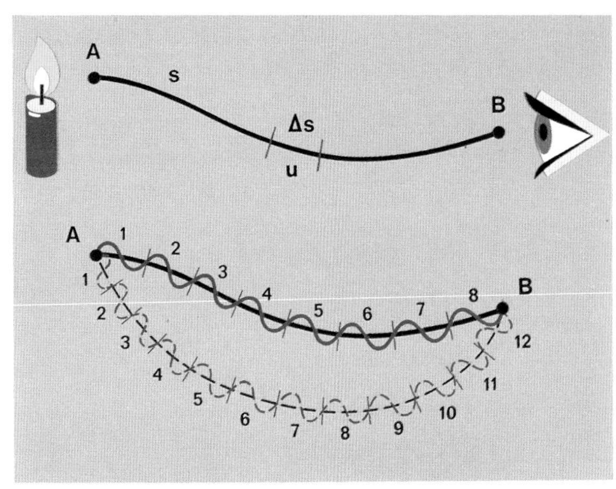

Path of light from A to B through a medium of variable refractive index. This path contains the minimum number (N=8) of wavelengths. Any other path (e.g. the dashed one, N=12), including the straight line from A to B, contains a larger number.

is a minimum, that is, the path that contains the smallest number of wavelengths, as indicated in the sketch. It has to be kept in mind that these wavelengths may be different, depending on how the index of refraction varies in the medium that is transversed by the light ray. In a medium of constant refractive index the light will travel in a straight line. Note that Fermat's principle itself has nothing to do with wavelengths. Indeed, it is doubtful whether Fermat had ever heard of the wave theory of light, as Christian Huyghens's (1629–1695) famous book *Traité de la Lumière* was published only in 1690, many years after Fermat's death.

Following the fundamental discoveries of Max Planck (1858–1947) and Albert Einstein (1879–1955), the French physicist Louis de Broglie (1892–1981) postulated in 1923 that every particle in motion, for example, an electron, is accompanied by a 'wave' whose wavelength Λ becomes shorter, the faster the particle is traveling, that is, the greater its velocity v, its momentum $p = m_e v$ and, accordingly, its kinetic energy $E = m_e v^2/2$, where m_e is the electron mass.

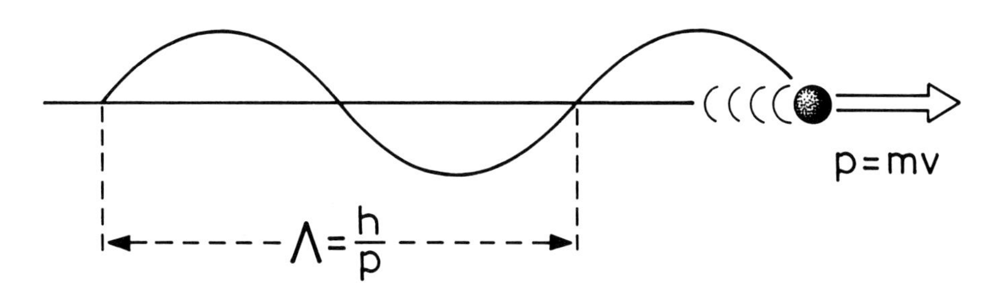

The exact relationship is given by the de Broglie equation $\Lambda = h/p$, in which h is Planck's constant. This naturally raised the question, whether one could proceed in exactly the opposite way, that is, describe the motion of such particles, and especially of electrons, by a type of theory that had been up till then reserved for describing waves. Indeed, only three years later, in 1926, the Austrian physicist Erwin Schrödinger (1887–1961) succeeded in achieving this and thus became the father of the so-called wave mechanics, which is today the basis for the description of electrons in atoms and molecules. It is interesting that the derivation of wave mechanics involves the assumption that all particles move in such a way that their path between two points, measured in de Broglie wavelengths, is again a minimum, harking back to the Fermat principle mentioned above. The connection can be illustrated in terms of another principle, the Principle of Least Action, first formulated by Pierre Louis Moreau de Maupertuis (1698–1759), according to which the path followed by any moving object is the one for which the sum of all $p(\Delta s)$ contributions is a minimum – where $p = mv$ is the momentum of the object as it traverses the path element Δs.

In terms of the de Broglie relationship, the quantity $\Sigma p\Delta s$ is equivalent to $h\Sigma(\Delta s/\Lambda)$ so that minimization of the former – the Maupertuis condition – is equivalent to minimization of $\Sigma(\Delta s/\Lambda)$, which is just the same condition as we found for light. Thus, as Schrödinger showed, the known formalism for the propagation of light waves can be extended to cover the motion of all particles; particles behave like waves, and waves behave like particles.

In addition to this principle, electrons have to obey certain 'traffic rules' when several of them co-exist in a confined space, as in an atom or a molecule. One of these rules has a great similarity to the behavior of the angel described by St. Thomas Aquinas and quoted in the chapter heading, but others are far beyond the scope of any analogy taken from classical physics or theology.

The most primitive model for electrons in a confined space, such as a molecule, is simply to confine them in a box of an appropriate size[39], as illustrated in the adjacent Figure with considerable poetic license, for the electrons of naphthalene – naphthalene (**17**), $C_{10}H_8$, contains not merely three but 68 electrons!

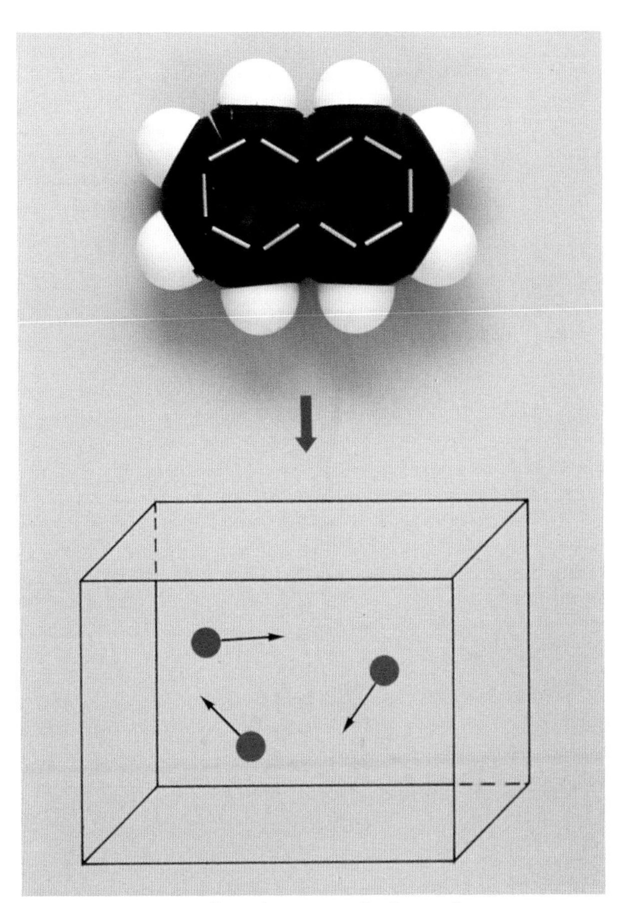

Naphthalene molecule regarded as a box containing electrons.

How do electrons respond when the space available to them is restricted? To provide at least a qualitative illustration, we first consider the very simplest case of a single electron moving backwards and forwards in a one-dimensional box. If this motion were to proceed according to the rules of classical mechanics, the electron would traverse the box with constant speed, bounce over and over against the walls of the box, and be reflected back each time to travel in the opposite direction but with always the same speed.

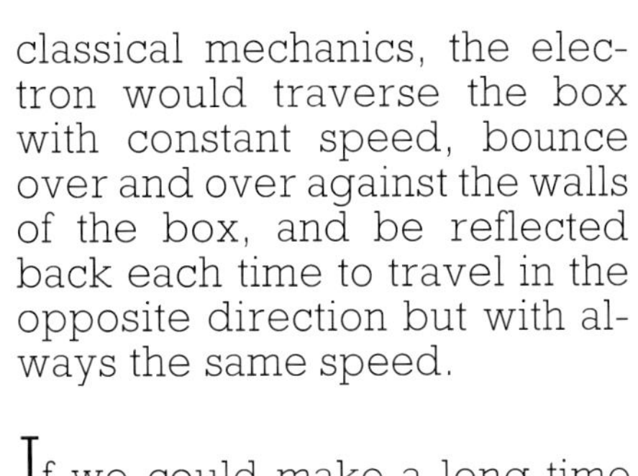

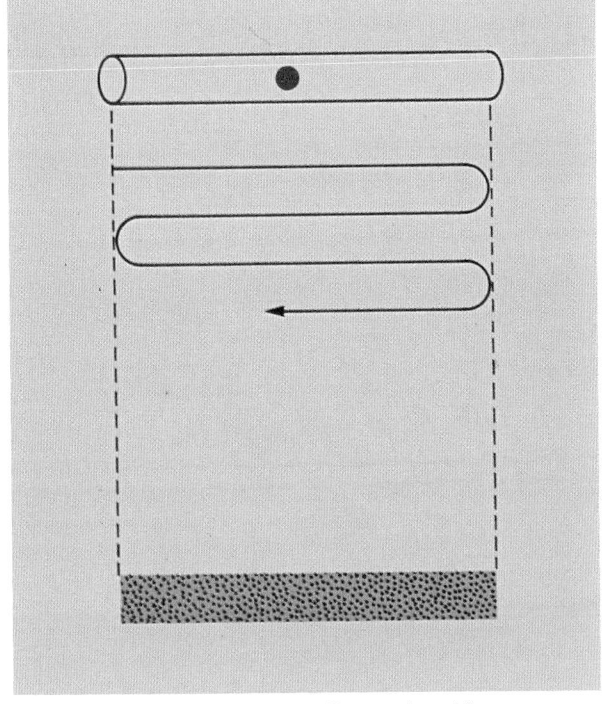

Classical electron in a one-dimensional box.

If we could make a long time exposure of such a classical electron, we would obtain a picture like the one shown in the adjacent Figure: the electron would be uniformly smeared over the entire region, or, to express this in

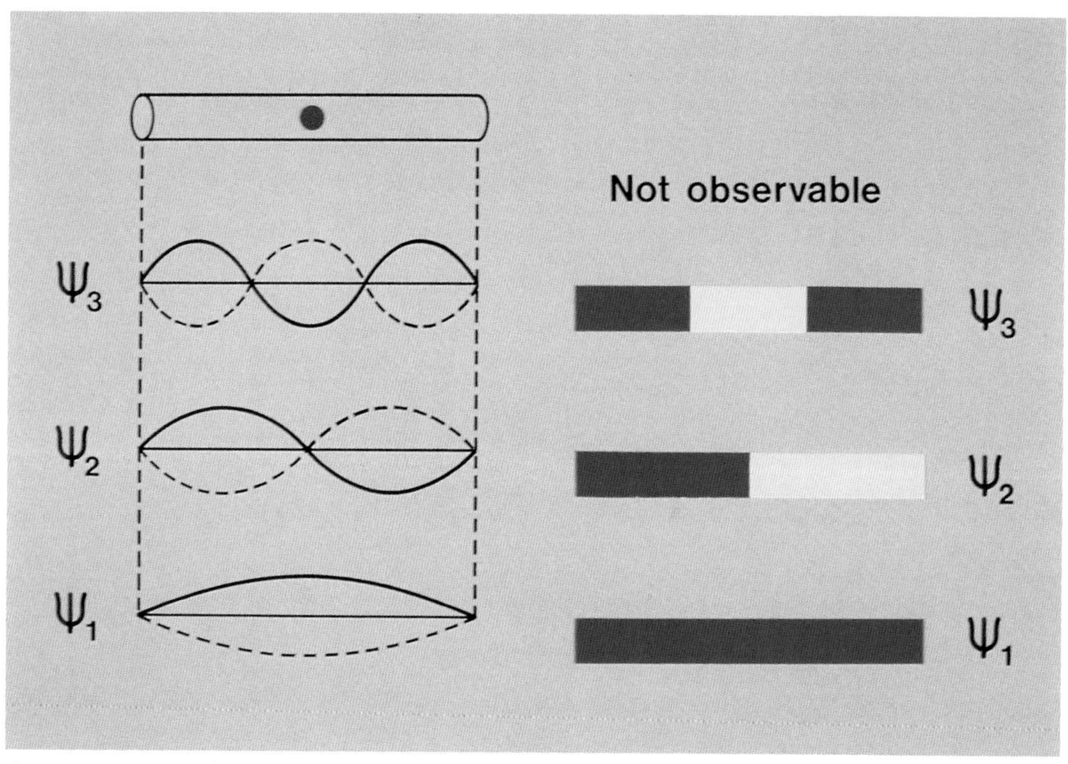

Quantum-mechanical electron in a one-dimensional box.

somewhat different language, the probability of finding the electron at some arbitrary time would be the same for all points in the box. Obviously, this result is completely independent of how fast the electron moves; anyway, our model imposes no limitations on its speed and hence its kinetic energy. In the wave mechanical description, this is quite incorrect. Wave mechanics requires that the wave that accompanies the electron must fit exactly into the box, as shown in the Figure above, for the three longest waves.

More exactly, as the Figure shows, an integral number of half-wavelengths must fit in the box, so if L is the length of the box then $n(\Lambda/2) = L$ where n takes the values 1, 2, 3,... etc. This kind of relationship is known as a quantization condition, the number n as a quantum number. The relationship in question means that only quite definite electron velocities and hence energies are permitted. One says that the energy is quantized, because it is only possible to pass from one state to another if an exactly defined quantum of energy is added to or removed from the system – here the electron in its one-dimensional box.

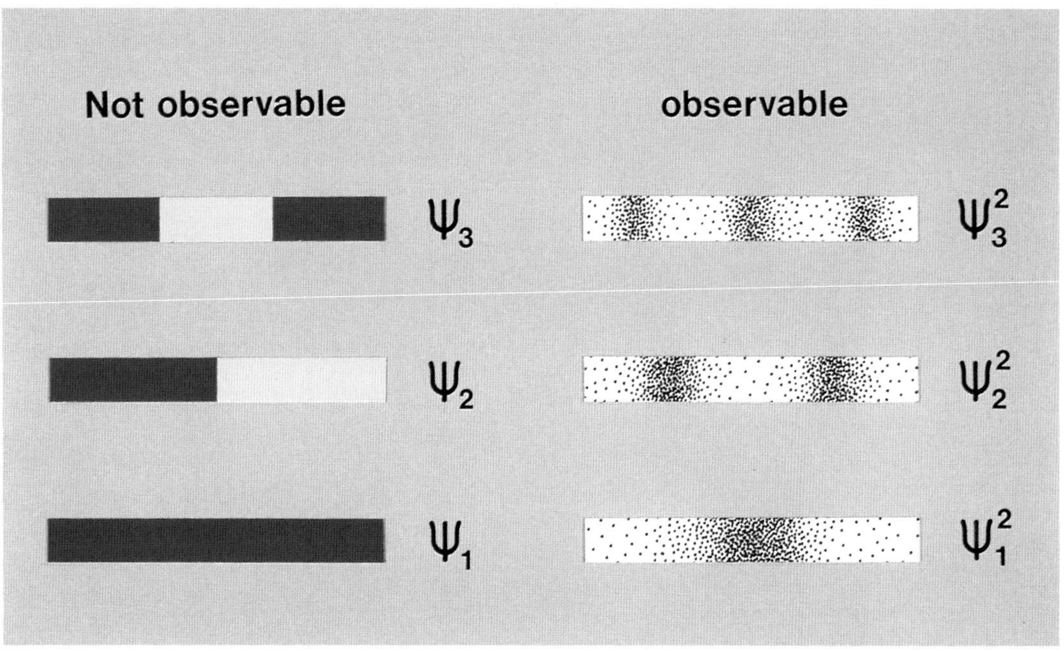

Schematic representation of wave functions and their squares.

The wave function in itself has no physical significance. First of all, it must be stated that there is no way of telling whether, at any given moment, the wave corresponds to the full or to the dotted line in the diagram. It may, therefore, be advantageous to symbolize the wave functions as shown on the right side of the above diagram, where the opposite deflections of the wave are indicated by the contrasting colors, blue and yellow.

In contrast to the classical behavior described earlier, the wave mechanical electron is not uniformly distributed over the space available to it. The probability of finding it at any given point is proportional to the square of the amplitude of the wave at that point. In contrast to the wave itself, the resulting electron 'density' is accessible to direct measurement, that is, it is an experimentally measurable quantity. Our simplified symbols for the wave and for the resulting electron density are shown face to face in the diagram above for the three states of lowest energy.

To recapitulate: on the left are the symbols for the wave functions, where the contrasting colors signify opposite deflections; and on the right are the corresponding electron densities or probabilities, smeared over the length of the box. The latter are observable, the wave functions are not. It is customary to name the wave functions as ψ_1, ψ_2, ψ_3....in order of increasing energy (which is the same as the order of decreasing wavelength since it can be shown that the energy of each such wave function is directly proportional to the square of the quantum number n, or, what is the same, inversely proportional to the square of the wavelength). The corresponding electron densities are given by the squares of the wave functions $(\psi_1)^2$, $(\psi_2)^2$, $(\psi_3)^2$, etc.

If we have only one electron, the state of lowest energy is obtained when the electron is described by the wave function of lowest energy ψ_1. We say that the electron occupies the ψ_1 orbital. With many electrons, one might have thought that the state of lowest energy is obtained by putting them all into the ψ_1 orbital. This is forbidden, however, by one of the fundamental rules that limit the motion of electrons: a given wave function or orbital can accommodate at most two electrons (and only then if the two have opposite spin). This is a consequence of a deep symmetry principle (too deep for us to go into here) first stated by the Zurich physicist Wolfgang Pauli (1900–1958). With many electrons, the state of lowest energy is obtained by filling the orbitals in order of increasing energy, two at a time, until all the electrons are accommodated.

Our one-dimensional box, for the time being without its electron, is symmetric. If rotated 180° about an axis perpendicular to the paper and passing through the mid-point of the box, it is brought into coincidence with itself; C_2 is a symmetry operation of the empty box. According to physics, the observable electron density $(\psi)^2$ produced by the electron moving in the box must have just the same symmetry as the box itself. This is seen to be the case for the diagrams in the Figure on the previous page. More generally, this means that electrons confined within a small space by a potential, in our case the box, will have the same probability of occurring at symmetry-equivalent points, so that at these points the electron densities will be equal. The physicist would say that the potential impresses its symmetry on the electron density. All this seems, and is, eminently reasonable.

A glance at the preceding Figure shows that the wave functions have a more complicated symmetry behavior. Some, such as ψ_1 and ψ_3, are indeed brought into coincidence with themselves by the rotation in question, but not others, ψ_2 for example. Here the two colors are interchanged by the C_2 operation, which means that the whole wave function changes its sign. Where there was a peak is now a valley, and vice versa. If we were to keep strictly to our previous definitions, then ψ_2 would not be symmetric with respect to C_2, but everyone will agree that this 'unsymmetric' behavior is quite different to the kind that pertains when we rotate a misshapen potato through 180°. The latter is unsymmetric with respect to C_2 in the true sense of the word. When only the sign of the wave changes, as for ψ_2 under the C_2 rotation, we speak of 'antisymmetry'. We thus see that our wave functions for an electron in a one-dimensional box are either symmetric (ψ_1, ψ_3, etc.) or antisymmetric (ψ_2, ψ_4, etc.). Such symmetry-antisymmetry relations are to be found quite generally for the wave functions of the electrons in molecules, whereby a whole variety of different kinds of symmetric-antisymmetric functions are possible, depending on the various symmetry operations that a molecule may possess. It is clear that a considerable extension and enrichment of symmetry theory is involved here.

As a simple example of this new kind of symmetry behavior, consider the motion of an electron in a rectangular two-dimensional box, as indicated in the next Figure. A classical particle would bounce around the box like a billiard ball, and, in particular, it would be confined to the inside of the box since it could never escape outside the reflecting walls. The symmetry of the box, that is, of the potential that limits the motion of the particle, is here defined by the following symmetry operations: a 180° rotation (C_2) about an axis through the midpoint of the box and perpendicular to its plane, and two reflections, σ_{xz}

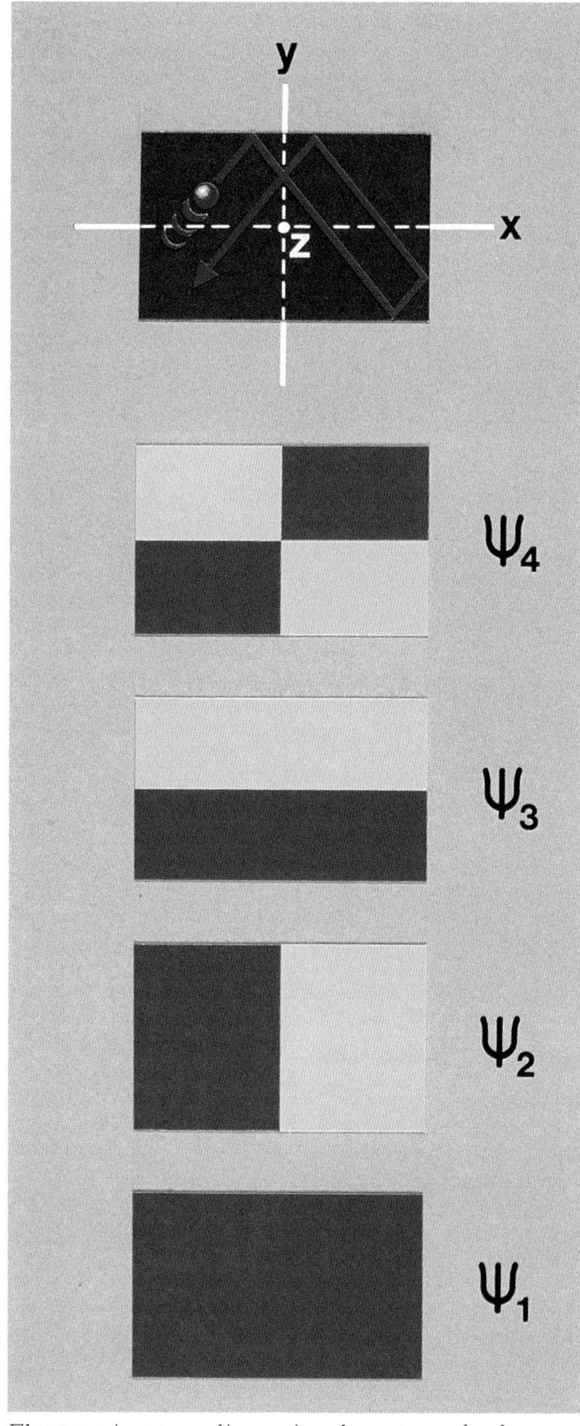

Electron in a two-dimensional, rectangular box, with schematic representations of the four wave functions lowest in energy.

and σ_{yz}, across the planes containing the x,z and y,z axes, respectively.

Here too we have a quantization condition that specifies the number of nodes in the allowed wave functions. To represent the alternation of sign in different regions of the box, we again use contrasting colors without specifying which color corresponds to positive and which to negative values. The adjacent Figure shows four such wave functions, labelled as ψ_1, ψ_2, ψ_3, ψ_4. While ψ_1 has the same sign throughout the whole area of the box, ψ_2 has opposite signs on opposite sides of the y,z reflection plane, ψ_3 has opposite signs on opposite sides of the x,z reflection plane, and ψ_4 shows a chessboard-like pattern. Each wave function is either transformed into itself or changes sign by the action of each of four symmetry operations (we include here the identity operation for completeness). The wave functions can thus be characterized by specifying this behavior, whereby we denote the first case as +1 and the second as −1. The behavior of the four wave functions ψ_1, ψ_2, ψ_3, ψ_4 is then as given in the following Table:

	I	C_2	σ_{xz}	σ_{yz}
ψ_1	1	1	1	1
ψ_2	1	−1	1	−1
ψ_3	1	−1	−1	1
ψ_4	1	1	−1	−1

These combinations, known as irreducible representations, are the only ones possible for this group of symmetry operations, and it follows that any acceptable wave function, regardless of the number and type of nodes, must transform according to one of them. Alternatively, for those who like to use letters instead of numbers, we can use the symbols *s* = symmetric and *as* = antisymmetric, in which case the Table looks like:

	I	C_2	σ_{xz}	σ_{yz}
ψ_1	*s*	*s*	*s*	*s*
ψ_2	*s*	*as*	*s*	*as*
ψ_3	*s*	*as*	*as*	*s*
ψ_4	*s*	*s*	*as*	*as*

The extension to three dimensions is obvious. To a first approximation, a lump of metal can be considered as a box containing a large number of electrons that are constrained to be within the box by a potential that is uniform throughout the box but rises steeply at the surface boundary. For a piece of metal of macroscopic dimensions, the allowed wavelengths Λ are so long and the energy levels so closely spaced that they

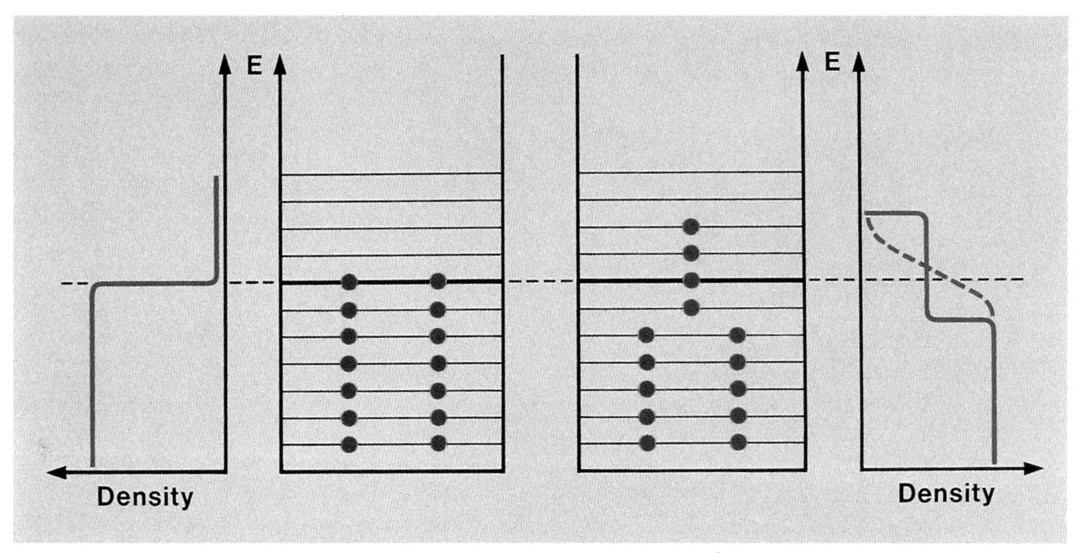

Energy levels occupied at absolute zero and at a higher temperature.

form an almost continuous range of energies. But the number of electrons is also enormously large, of the order of 10^{22} per cubic centimeter. According to classical mechanics, they could all have the same energy, and, at the absolute zero of temperature, they could all occupy the lowest energy level. According to quantum mechanics, at absolute zero all the orbitals up to a certain energy level will be occupied by pairs of electrons with antiparallel spin, while orbitals having energies greater than this will be empty. However, as the temperature rises, electrons having energies close to this level can gain thermal energy, which allows them to move into the higher, unoccupied levels, as indicated in the Figure above for the case of a one-dimensional box. The electrons in the singly occupied orbitals can move freely within the metal, and under the influence of an external potential, they are responsible for conducting the electric current.

With the help of this kind of extended symmetry consideration and on the basis of wave functions that represent the motion of electrons in molecules, the chemist has learned to make predictions about the course of reactions and about the structures of the products. Some of these predictions would not be possible in other ways. Even the rules governing the local signs of the wave functions can enable far-reaching conclusions to be drawn in a relatively simple way. This is illustrated by an analogy, which may admittedly not quite hit the mark.

We have an 8 × 8 frame with two squares missing at opposite ends of a diagonal, and there are 31 dominoes, each of which can cover two squares. The question to be answered is: can one completely cover all 62 squares with the 31 dominoes?

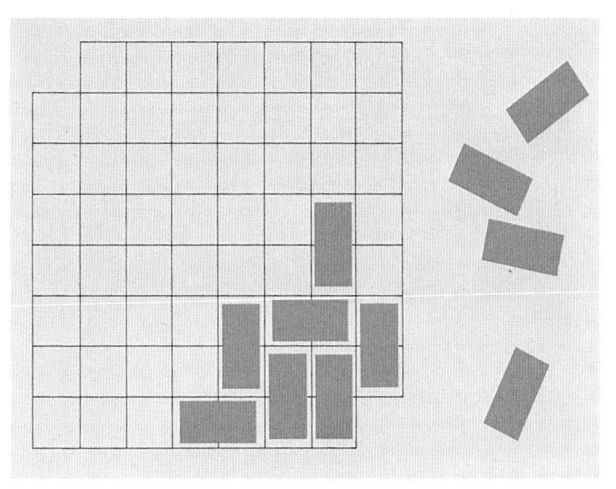

Can 31 dominos cover the 62 squares ?

Expressing the problem in a chemical context, one could say that the empty frame and the dominoes represent the 'educts', and the question can now be rephrased in terms of whether they can react to give the desired 'product', the completely covered frame, or not. Making use only of these components and their rather obvious symmetry properties, the question may seem to be not easily answerable. In particular, trial and error leads nowhere.

The situation is completely altered when one considers the underlying 'wave functions' of the frame, which is then recognized to be nothing more than a chessboard with the squares at two opposite corners missing. Consideration of the colors of the individual squares shows immediately that our problem can be solved in a trivial way. The answer is 'no'. Squares at opposite corners of a chessboard have the same color or 'sign'; in the next Figure they are both white, so the frame consists of 30 white and 32 black squares. Since each domino must cover two squares of opposite color or 'sign', two black squares, which cannot adjoin one another and hence cannot be covered with the last domino, must always be left over. There is no way the conditions can be satisfied, meaning that our 'educts' cannot possibly lead to the desired 'product'. Admittedly, the analogy is somewhat far-fetched, but it strikes at the heart of the matter, as the following example should show.

A preliminary remark is necessary here: the simplest molecule is the hydrogen molecule H_2, in which two hydrogen

The answer is NO !

nuclei are held together by two electrons. The formation of this molecule can be pictured in the following primitive, but for our purposes adequate, manner. The left-hand side of the Figure below shows the two hydrogen atoms, or, more precisely, the symbolic representation of the wave functions for the two electrons that move round their respective nuclei. The two horizontal lines indicate the energy of these electrons on the energy scale shown on the left.

To form the molecule the two atoms must be brought together, and this can be done so that their wave functions have either the same sign or opposite signs. When the wave functions overlap, the combination with the same sign yields a

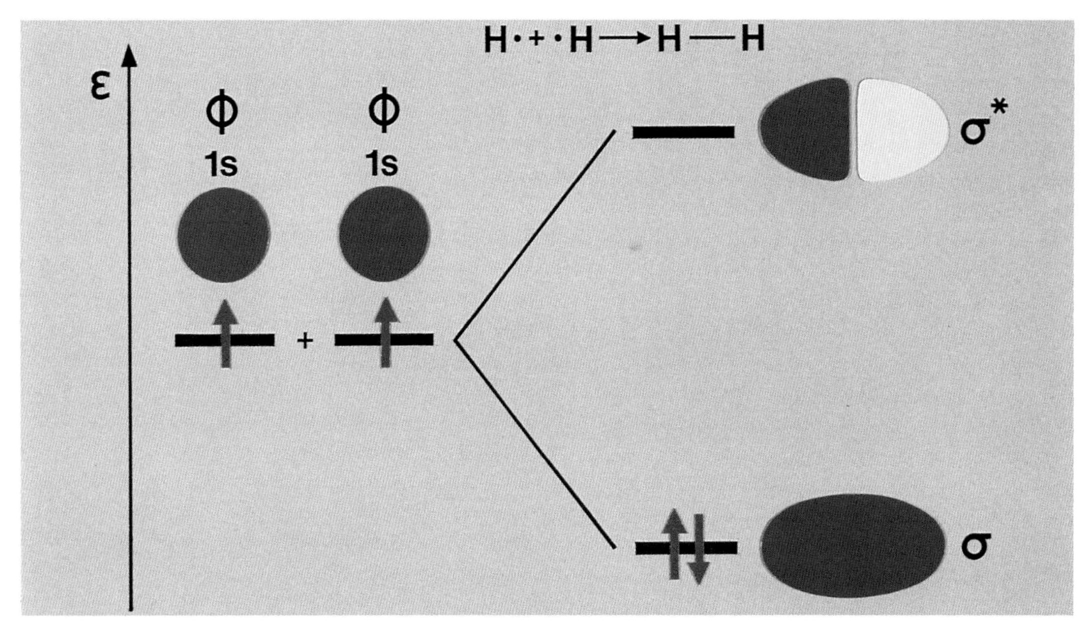

When two hydrogen atoms approach one another, their 1s orbitals overlap to give a bonding and an antibonding combination.

resultant wave function corresponding to a lower-lying, more stable energy level; conversely, the combination with opposite signs gives a higher-lying, less stable energy level. In the molecule, the symmetric, lower-lying level accommodates the two electrons. The antisymmetric combination, which has one node and is hence of higher energy (analogous to the energies of the wave functions for the electron in a box), corresponds to a destabilization relative to the separated atoms. Reducing this to a simple rule, overlapping of two wave functions of the same sign means bonding, overlapping of two of opposite sign means repulsion. This rule enables us to provide simple answers to complicated questions – like our chessboard example.

Before we proceed further, a few words about the electron wave functions in atoms and simple molecules are called for. The above description of the bond in the hydrogen molecule H_2 starts out from the wave functions Φ_{1s} for the motion of the electrons in the two separated hydrogen atoms. These functions were spherically symmetric and had the same sign throughout the space occupied by the electrons in question, similar to the ψ_1 functions of our one- and two-dimensional boxes. Again analogous to the boxes, the hydrogen atom has states of higher energy described by wave functions containing nodes – the greater the number of nodes, the higher the energy. The smallest possible energy increment thus leads to a wave function with one node, for example, to a wave function similar to ψ_2 with one horizontal plane of antisymmetry.

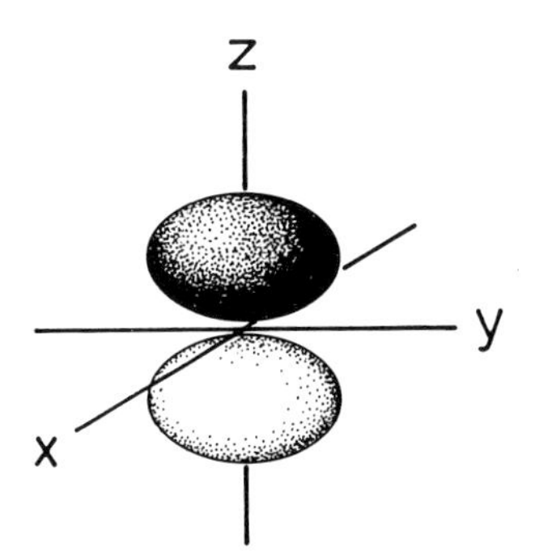

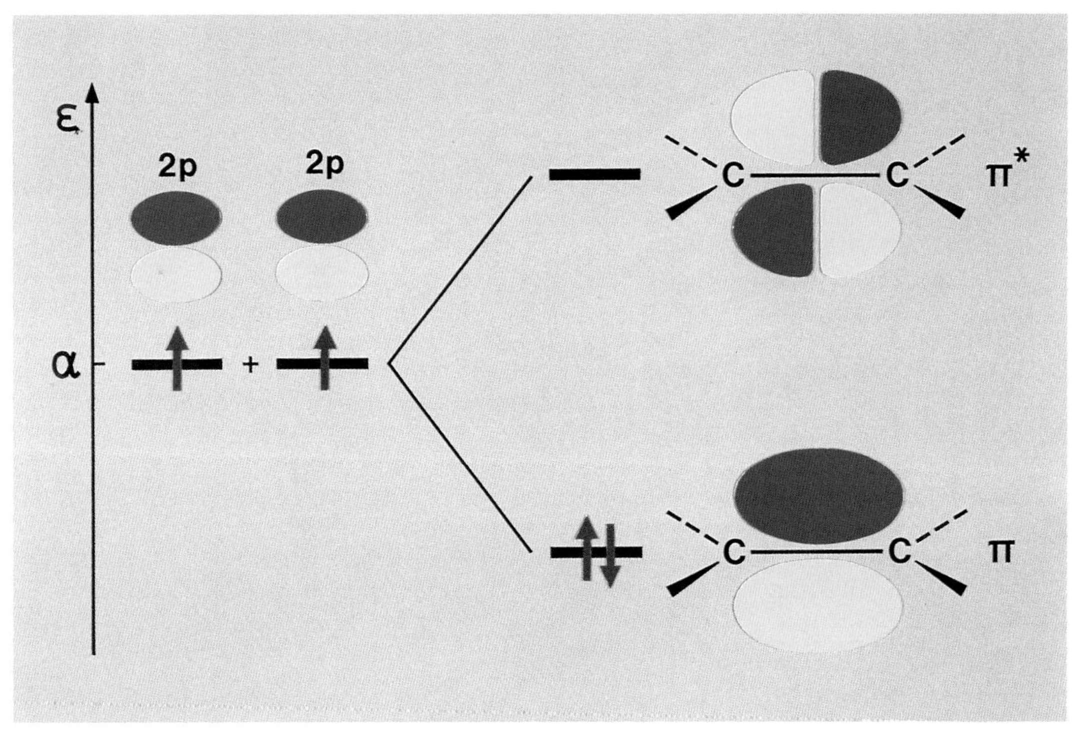

Formation of the bonding (π) and antibonding (π) molecular orbitals of ethylene from a pair of atomic p orbitals.*

Wave functions that describe the motion of an electron in the field of an atomic nucleus are called atomic orbitals. Atomic orbitals may be spherically symmetrical (s orbitals) or they may have a plane of symmetry (p orbitals) or more complicated symmetry behavior (d and f orbitals). The letters that describe them derive from the early history of atomic spectroscopy and were taken over into quantum mechanics when it became possible to explain patterns of spectral lines by quantum mechanical theories.

Molecular orbitals may be obtained by combining atomic orbitals belonging to different atoms. Just as we can describe the bonding orbital of the H_2 molecule as a combination of two s atomic orbitals, we can also combine a pair of p orbitals to give a bonding orbital. Such orbitals have a special role because the electrons whose motion is described by them are of relatively high energy and are important in determining the chemical and physical properties of molecules. The Figure above shows the combination of two atomic p orbitals to give a

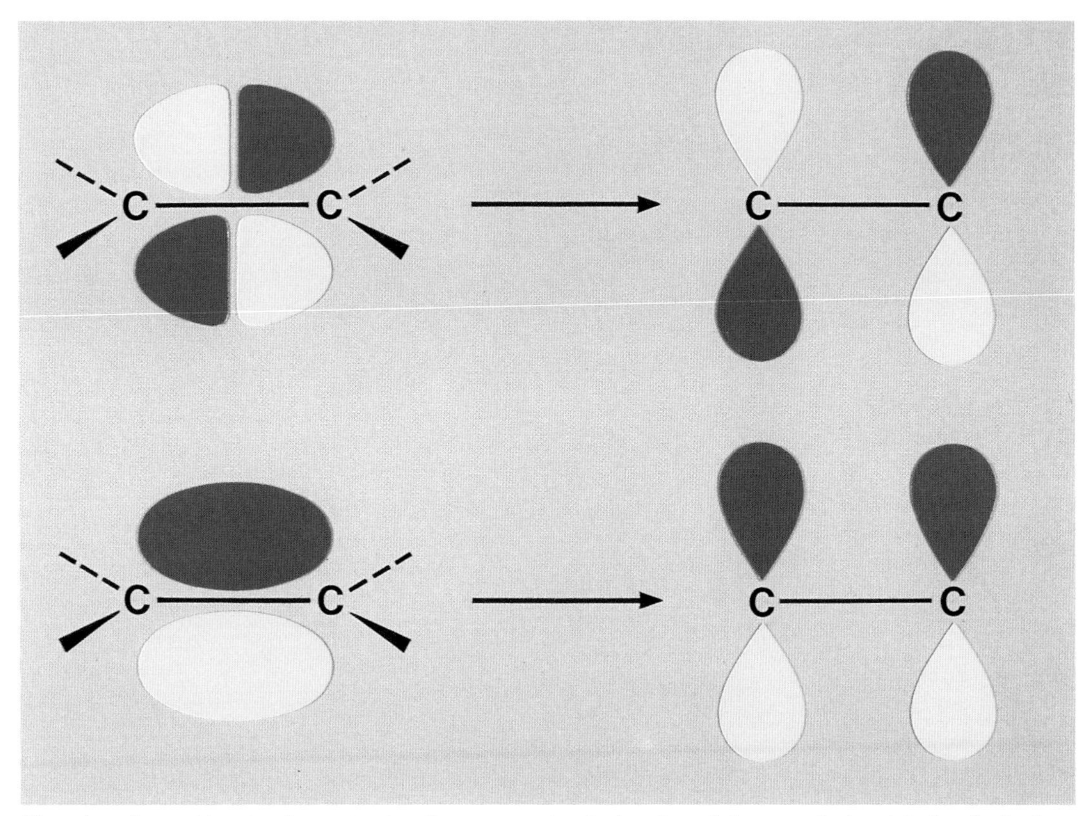

Shorthand notation to characterize the symmetry behavior of the π and π^* orbitals of ethylene.

bonding orbital of the ethylene molecule $H_2C{=}CH_2$ (compare with the corresponding Figure on page 138 for the H_2 molecule). This type of bonding orbital obtained by the positive combination of the two p orbitals is known as a π orbital. The negative combination, the π^* orbital has one additional node and thus has a higher energy than the π orbital.

Ethylene is a symmetrical molecule, containing the three reflection planes σ_{xy}, σ_{yz}, and σ_{zx} among its symmetry elements. If the molecule is reflected across any of these planes it is transformed into itself. The symmetry behavior of the π and π^* molecular orbitals is more complicated. The former is symmetric with respect to σ_{yz} and σ_{zx} but antisymmetric with respect to σ_{xy}, while the latter is symmetric only with respect to σ_{yz}, antisymmetric with respect to the two others. Note, however, that the corresponding electron densities, given by the squares of the orbitals, show the full symmetry of the ethylene molecule.

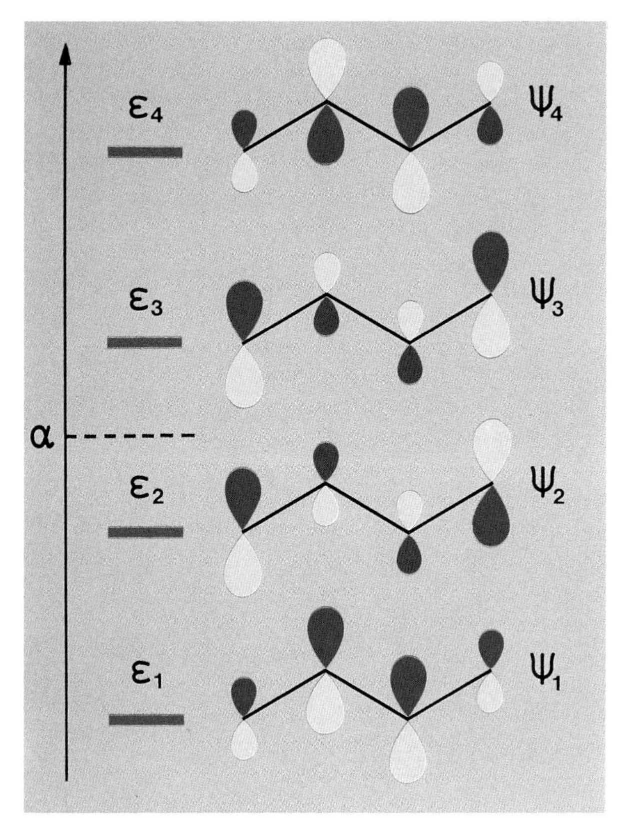

Molecular π orbitals of butadiene.

Chemists have become accustomed to a kind of shorthand in which the symmetry behavior of molecular orbitals is depicted by showing the individual atomic orbitals, each carrying a relative sign corresponding to its contribution to the combined orbital, as indicated in the previous Figure for ethylene.

The adjacent Figure shows in the same way the combinations that can be made from the atomic p orbitals of the four carbon atoms of the butadiene molecule $H_2C{=}CH{-}CH{=}CH_2$.

We now consider a simple example where the symmetry properties of such molecular orbitals can be used to make predictions about the specific outcome of a chemical process. The molecules naphthalene (**17**) and tetracyanoethylene (**43**) have the ability to combine together in a so-called molecular complex. In this complex, the planar naphthalene and tetracyanoethylene molecules stack one above the other, but it is a priori not obvious what relative position they should adopt. In the next Figure, two possibilities are indicated, both corresponding to arrangements that have the highest possible symmetry and might therefore come into question if Nature should prefer the most aesthetically pleasing solution.

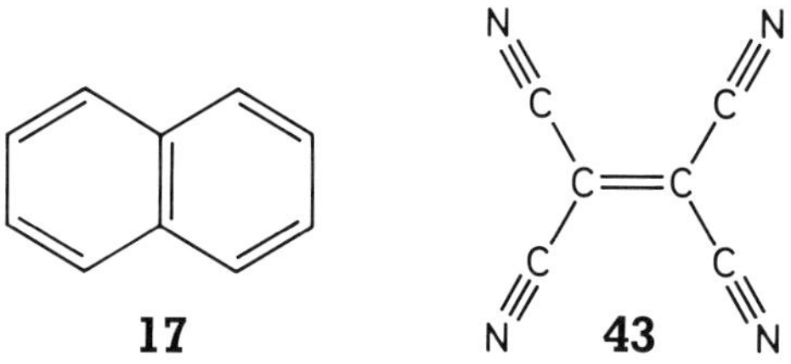

17 **43**

Two possible symmetric arrangements of a naphthalene tetracyanoethylene complex.

On the basis of symmetry considerations and mathematical calculations that are of no interest here, the signs of the wave functions of the electrons responsible for the bond between the two molecules can be calculated. The result of the calculation is shown graphically in the Figure below, which shows the relevant molecular π orbitals in the same shorthand notation as used on the opposite page for butadiene, with the only difference that we are now looking from one side of the plane of the molecule.

According to the rule mentioned above, the two molecules should be superimposed in such a way as to secure the best possible agreement between the signs of adjacent atoms. This occurs for the arrangement shown in the Figure on the next page. For the sake of clarity, we have introduced a slight inconsistency into the picture. Each of the two molecular orbitals is antisymmetric with respect to the plane in which the atoms lie. The diagram must be understood to show how the signs of the two molecular orbitals combine in the region between the molecules to produce the 'in phase' overlapping of the atomic orbitals.

It may seem surprising that this is not one of the beautifully symmetric arrangements but rather one where the tetracyanoethylene sits over one side of the naphthalene molecule. An X-ray analysis of the crystalline complex shows that the low-symmetry arrangement is indeed realized. Obviously, Nature follows symmetry criteria that represent an extension and enhancement of the purely intuitive concept. Although everything is therefore much more complicated than Kepler ever imagined, he would nevertheless have been very satisfied to

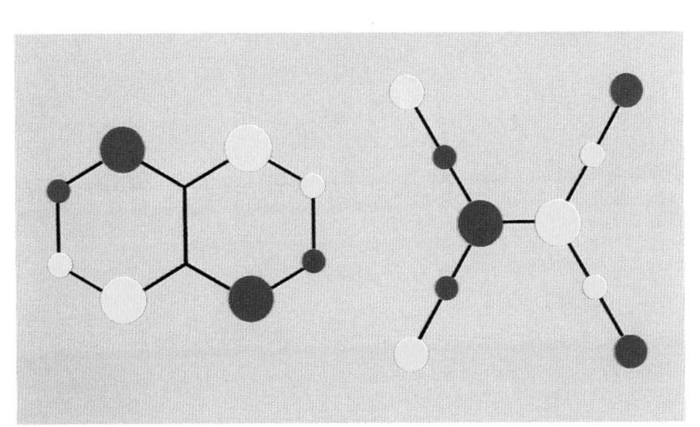

Signs of the interacting orbitals of naphtalene and tetracyanoethylene.

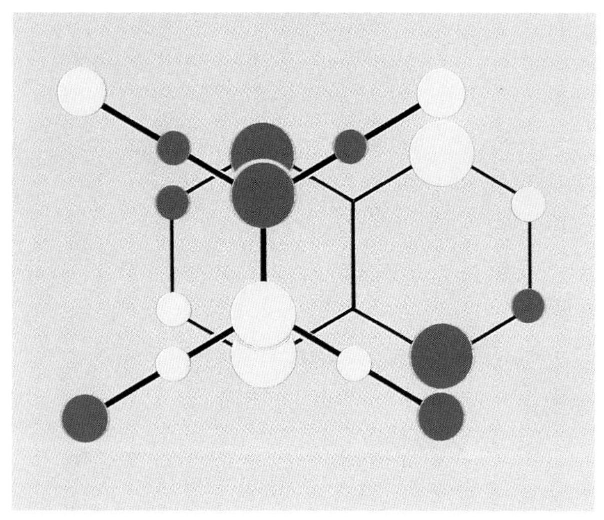

Optimal overlapping of the orbitals shown in the previous Figure.

know that symmetry largely determines the behavior and properties of molecules.

Quite generally, the overlapping of atomic orbitals with the same sign leads to a stabilization of the system, in other words to bonding, while overlapping of orbitals with opposite sign leads to destabilization. The systematic application of this principle to explain chemical reactivity was initiated by the American chemists Robert Burns Woodward (1917–1979) and Roald Hoffmann. Their principle of the Conservation of Orbital Symmetry[40] classified certain types of reactions as being symmetry-allowed or symmetry-forbidden in terms of a few simple rules based on the symmetry behavior of the orbitals involved. Symmetry-allowed reactions proceed smoothly from reactants to products, whereas symmetry-forbidden ones either proceed by roundabout routes involving intermediates or do not occur at all. This conceptual approach was so far reaching and successful that it opened the eyes of a generation of practical chemists to the potentialities of simple theoretical models, especially those derived from molecular orbital theory. After many years of comparative neglect, molecular orbitals suddenly became fashionable.

Our final example illustrates how orbital symmetry considerations can be used to predict the course of a chemical reaction. Say we start with an unsymmetrically 1,4-disubstituted butadiene molecule; call one of the substituents R, the other Q. If the two outer carbon atoms are now joined together to make a four-membered ring, two isomeric products are possible: a *cis*-disubstituted cyclobutene with the R and Q groups on the same side of the ring plane, or a *trans*-disubstituted cyclobutene with the two groups on opposite sides, as shown in the next Figure.

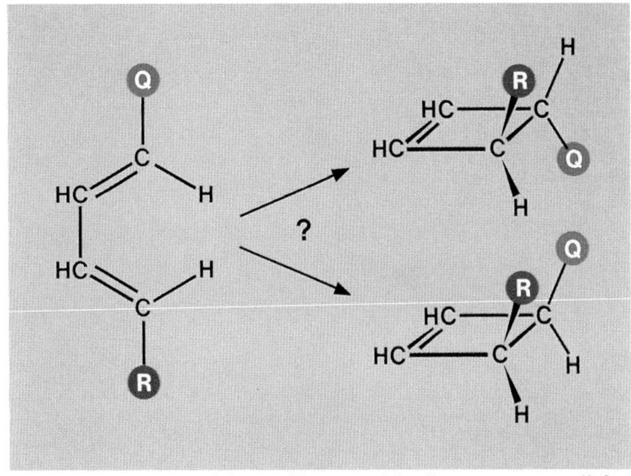

1,4-Disubstituted butadiene with the two possible isomeric cyclization products.

We now make use of a heuristically very successful idea introduced by the Japanese theoretical chemist Kenichi Fukui, namely, the assumption that the chemical behavior of a molecule is determined primarily by its highest-lying occupied orbital, the one in which the most loosely bound electrons are found. This idea had previously been applied to atoms: many properties of a given atom can be correlated with its ionization energy, that is, the energy required to remove one of the 'outer' or 'valence' electrons. In its extension to molecules, it implies that the symmetry properties of the high-lying orbital, the 'frontier' orbital, interpreted according to the Woodward-Hoffmann rules, are decisive in determining the course of certain chemical reactions.

The π molecular orbitals of butadiene are shown in the Figure on page 142. Each of the carbon atoms contributes one electron to the π system, so there are four electrons to be accommodated, and since only two electrons may be housed in the same orbital, the state of lowest energy is the one in which the two lowest orbitals ψ_1 and ψ_2 each contain two electrons. The higher of these, ψ_2, is then the highest occupied orbital, so this is the frontier orbital whose symmetry properties should determine the course of the ring-closure reaction, at least as long as the molecule is supposed to be in its electronic ground state.

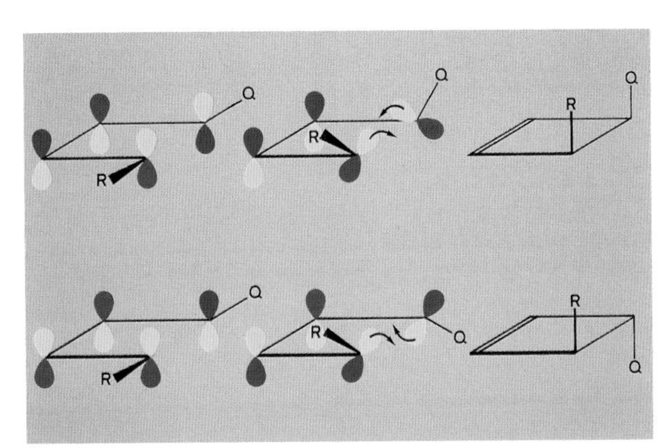

Conrotatory and disrotatory motions leading to two different cyclization products of 1,4-disubstituted butadiene.

On the left of the adjacent Figure is shown the disubstituted butadiene molecule in the conformation in which the atoms can be expected to lie before the ring-closure reaction occurs. On the lower left we see the frontier orbital ψ_2, which is now supposed to determine the further course of the reaction. According to the Woodward-Hoffmann rules, in order to form the new bond the orbital components at the two ends of the butadiene fragment must come together so that lobes of the same sign (here of the same color) match. This is possible only if the two end groups are rotated in the same sense, i.e., both clockwise or both anti-clockwise (the term 'conrotatory' is used to describe this type of coupled motion). The inevitable result is that the two groups R and Q end up on opposite sides of the ring plane. Thus, as long as the ring-closure reaction proceeds through the electronic ground state of the reacting molecule, we expect to obtain the *trans*-isomer.

But not all chemical reactions proceed through the electronic ground states of the reactants. Some reactions – photochemical reactions – are promoted or accelerated by light: the miracle of photosynthesis, whereby water and carbon dioxide are converted into the molecules of life by the action of sunlight, is the most important example, while the fading of the dyes contained in curtains and other fabrics, or the changes that take place in sunburned skin, are more trivial everyday examples. The first step in such chemical changes is the absorption of light, or, more exactly, the absorption of a photon or energy quantum by a molecule, which is thereby converted into a state of higher energy. If the photon energy is small, the excess energy merely goes into the molecular vibrations or rotations, but if it is large enough the molecule can be excited from its electronic ground state into an electronically excited state. Such a change typically involves the promotion of an electron from a low-lying occupied orbital to a higher-lying unoccupied one.

In our butadiene example, light absorption could lead to the promotion of an electron from the doubly occupied orbital ψ_2 to ψ_3, the lower of the two unoccupied orbitals shown in the Figure on page 142. In such a case, this electron would become the most loosely bound one, and hence it should be the symmetry properties of ψ_3 rather than those of ψ_2 which should be decisive in influencing the reactivity. As indicated in the upper part of the last Figure, in order to satisfy the overlapping requirement, the two end groups must now be rotated in opposite directions, i.e., one clockwise, the other anti-clockwise ('disrotatory' motion as distinct from 'conrotatory'). The result is that in the photochemical ring-closure reaction the two groups R and Q end up on the same side of the ring plane to give the *cis*-isomer.

At this point we have to confess that the above account of the cyclization of butadiene is not quite correct. In fact, the reaction as described above does not happen at all; it actually proceeds in the opposite direction, that is, it goes from the cyclobutene derivative to the open-chain butadiene. This means that the Woodward-Hoffmann rules should actually have been applied in the reverse direction, i.e., to predict the arrangement of the groups R and Q in the open-chain molecules, starting from their arrangement in the cyclic molecules. The experimental result is summarized below, and it is easily seen that this corresponds exactly to what would have been expected from the rules.

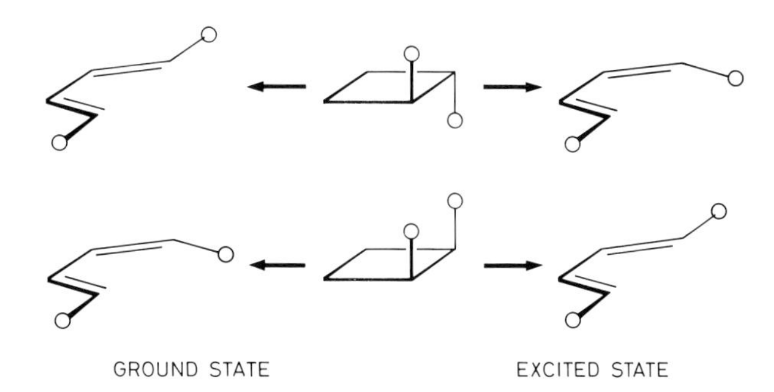

It goes without saying that these rules are enormously useful in planning the synthesis of complex molecules and in overcoming the stereochemical problems that arise there. It was, in fact, in the early stages of the synthesis of vitamin B_{12} that the effect of orbital symmetry rules was first discerned, and it was in its successful culmination by Woodward and by Albert Eschenmoser that the full power of these rules was made evident.

REFERENCES

1) Ferdinand Hodler (1853–1918), *Thunersee mit symmetrischer Spiegelung* (1909), Musée d'Art et d'Histoire, Geneva. We have been helped by Mr. Bernhard von Waldkirch, Kunsthaus Zürich, and Mr. Jura Brüschweiler in our efforts to trace the history of this painting.

2) Some of the qualities we associate with the periodic patterns of Maurits Cornelis Escher (1898–1972) are already present in designs by Koloman Moser (1868–1918). Werner Fenz, *Koloman Moser,* Residenz Verlag, Salzburg und Wien, 1984. The Escher pattern shown in the Figure is an adaptation taken from: D. Schattschneider and W. Walker, *M. C. Escher Kaleidocycles,* Ballantine Books, New York, 1977. We are grateful to Professor Doris Schattschneider for helpful correspondence and for permission to reproduce the pattern.

3) Raphael (Raffaello Santi, 1483–1520), Madonna alba, National Gallery, Washington DC.

4) J. Kepler, *De Nive Sexangula,* Gottfried Tampach, Frankfurt, 1611; *On the Six-Cornered Snowflake,* translation by C. Hardie, Clarendon Press, Oxford, 1966.

5) J. W. Shirley, *Thomas Harriot, Renaissance Scientist,* Clarendon Press, Oxford, 1974.

6) J. Dalton, *A New System of Chemical Philosophy,* Manchester, Vol. 1, 1808, 1810; Vol. 2, 1827; for details: F. Greenaway, *John Dalton and the Atom,* Cornell University Press, Ithaca, 1966.

7) W. A. Bentley and W. J. Humphreys, *Snow Crystals,* McGraw-Hill Book Comp. Inc., New York, 1931; reprinted: Dover Publications Inc., New York, 1962.

8) Wu-Yi Hsiang, *Sphere Packings and Spherical Geometry – Kepler's Conjecture and Beyond,* preprint, Center of Pure and Applied Mathematics, University of California, Berkeley, July 1991; for a review, see: I. Stewart, *Scientific American,* February 1992, p. 90, and references therein.

9) R. J. Bošković, *Theoria Philosophiae Naturalis,* Venice, 1763; L. L. Whyte, *Roger Joseph Boscovich,* George Allen & Unwin, London, 1961; Žarko Dadić, *Bošković,* Školska kujiga, Zagreb, 1987 (English translation by J. Paravic).

10) This proof was presented several years ago by Dr. Peter Murray-Rust to one of the authors. According to Martin Gardner (in M. Kline, Edit., *Mathematics; an Introduction to its Spirit and Use,* W.H. Freeman & Comp., San Francisco, 1978, p. 97) a group-theoretical proof has been published by: I. M. Yaglom, *Geometric Transformations,* Random House, New Mathematics Library, 1962, p. 93.

11) W.H. Wollaston, *On the Elementary Particles of Certain Crystals,* Bakerian Lecture, *Phil. Trans.* **1813**, *2,* 51; reprinted in: D.M. Knight, *Classical Scientific Papers – Chemistry,* Mills & Boon Ltd., London, 1968.

12) A. Kekulé, *Ann. Chem.* **1857**, *101,* 257; A.S. Couper, *Compt. rend.* **1858**, *46,* 1157; *idem, The London, Edinburgh and Dublin Philos. Mag. and J. Sci.* **1858**, *16,* 104; reprinted in Ostwald's *Klassiker der exakten Naturwissenschaften,* Nr. 183, W. Engelmann, Leipzig, 1911.

13) J.H. van't Hoff, *Voorstel tot uitbreiding der tegenwoordig in de scheikunde gebruikte structurformules in de ruimte, benevens een daarmee samenhangende opmerking omtrent het verband tusschen optisch actief vermogen en chemische constitutie van organische verbindingen,* Utrecht 1874; essentially the same idea was published by the French chemist Jules Achille Le Bel (1847–1930) about the same time: *Bull. Soc. Chim. (Paris)* **1874**, *22,* 337.

14) R. Anschütz, *August Kekulé,* Verlag Chemie GmbH., Berlin, 1929.

15) A. Kekulé, *Bull. Soc. Chim. Fr.* **1865**, *3,* 98; *Ann. Chem.* **1866**, *137,* 158.

16) P. Havrez, *Revue universelle des Mines, de la Métallurgie, des Travaux Publics, des Sciences et des Arts* **1865**, *18,* 318, 433; see also: O.B. Ramsay, *Chemistry* **1974**, *47,* 6.

17) The last of the four papers by Körner was published in *Gazz. Chim. Ital.* **1874**, *4,* 305, with references to the first three; reprint of all four papers: Ostwald's *Klassiker der exakten Naturwissenschaften,* Nr. 174, W. Engelmann, Leipzig, 1910; see also ref. 15.

18) J.M. McBride, *Completion of Koerner's Proof that the Hydrogens of Benzene are Homotopic. An Application of Group Theory, J. Am. Chem. Soc.* **1980**, *102,* 4134.

19) M.A. Gaudin, *L'Architecture du Monde des Atomes,* Paris, 1873.

20) R. Willstätter and E. Waser, *Ber. Dtsch. Chem. Ges.* **1911**, *44,* 3423. The compound was obtained in small quantities from the alkaloid pseudopelletierine. A detailed investigation of its properties became possible after W. Reppe prepared large quantities by high-pressure polymerization of acetylene (*Ann. Chem.* **1948**, *560,* 1).

21) A summary of the work of Erich Hückel can be found in: E. Hückel, *Grundzüge der Theorie ungesättigter und aromatischer Verbindungen,* Verlag Chemie GmbH, Berlin, 1938.

22) A. Baeyer, *Ber. Dtsch. Chem. Ges.* **1885**, *18,* 2269; H. Sachse, *ibid.* **1890**, *23,* 1363; E. Mohr, *J. Prakt. Chem.* **1918**, *98,* 315; *Ber. Dtsch. Chem. Ges.* **1922**, *55,* 230.

23) S. B. Hendricks and C. Billeke, *The Space-Group and Molecular Symmetry of β-Benzene Hexabromide and Hexachloride, J. Am. Chem. Soc.* **1926**, *48,* 3007; R. G. Dickinson and C. Billeke, *The Crystal Structure of Beta - Benzene Hexabromide and Hexachloride, J. Am. Chem. Soc.* **1928**, *50,* 764.

24) For a survey, see: H. Kroto, *Giant Fullerenes, Chem. Brit.* **1990**, 90; R. F. Curl and R. E. Smalley, *Fullerenes, Scient. Am.* **1991**, October, 32; and references given therein.

25) This method was developed by Z. Vager, E. P. Kanter, and their coworkers, *Phys. Rev. Lett.* **1986**, *57,* 2793; *J. Chem. Phys.* **1986**, *85,* 7487; for a review, see: G. Neubert, *Physik in unserer Zeit* **1989**, *20,* 153.

26) Shown are van't Hoff's original models, now at the Rijksmuseum voor de Geschiedenis der Natuurwetenschappen, Leiden.

27) From Friedrich Ebel's chapter, *Die Tetraedertheorie,* in Book 2, *Stereochemie der Kohlenstoffverbindungen,* of *Stereochemie* (ed. K. Freudenberg), Deuticke, Leipzig und Wien, 1933, p. 532.

28) A. Werner, *Beitrag zur Konstitution anorganischer Verbindungen, Z. anorg. Chem.* **1893**, *3,* 267; *Zur Kenntnis des asymmetrischen Kobaltatoms, Ber. Dtsch. Chem. Ges.* **1911**, *44,* 1887; reprinted in Ostwald's *Klassiker der exakten Naturwissenschaften,* Nr. 212, Akademische Verlagsgesellschaft, Leipzig, 1924.

29) Alexander Rich of the Massachusetts Institute of Technology has found that, under certain conditions, DNA forms a left-handed double helix which, of course, is *not* the mirror image of normal DNA.

30) Reproduced with permission from: R.E. Dickerson and I. Geis, *The Structure and Action of Proteins,* Harper & Row Publishers, New York, 1969.

31) The synthesis of a mirror-image protein, the protease from HIV virus, has just been reported by R. C. de L. Milton, S. C. F. Milton, and S. B. H. Kent, *Total Synthesis of a D-Enzyme: The Enantiomers of HIV-1 Protease Show Demonstration of Reciprocal Chiral Substrate Specificity, Science* **1992**, *256,* 1445.

32) P. A. M. Dirac, *Forms of Relativistic Dynamics, Rev. Mod. Phys.* **1949**, *21,* 392. See also the memoir by R.H. Dalitz and R. Peierls in *Biographical Memoirs of Fellows of the Royal Society* **1986**, *32,* 159.

33) E. Jandl, *Laut und Luise,* Luchterhand Literaturverlag, Frankfurt, 1990; drawing from *Nebelspalter.* We thank the Luchterhand Literaturverlag for permission to reprint the poem, and the Nebelspalter Verlag for permission to reproduce the drawing.

34) These platonic solids with Escher patterns have been constructed from the cut-outs designed by D. Schattschneider and W. Walker, *cf.* ref. 2.

35) H.E. Armstrong, *Poor Common Salt,* Letter to the Editor, *Nature (London)* **1927**, *120,* 478.

36) The authors have tried to trace the origin of this figure, unfortunately without success.

37) J. D. Dunitz and E.F. Meyer, *Structure Analysis of Nickel(II)-1,8,8,13,13-pentamethyl-5-cyano-*trans*-corrin chloride, Proc. R. Soc.* **1965**, *228A,* 324.

38) D. Shechtman, I. Blech, D. Gratias, and J. Cahn, *Phys. Rev. Lett.* **1984**, *17,* 1951; D. Gratias, *La Recherche* **1986**, *17,* 788; D.P. DiVincenco and P.J. Steinhardt, Eds., *Quasicrystals,* World Scientific, Teaneck N.J., 1991.

39) The 'electron-in-a-box' model for extended π-systems has been introduced and developed by Hans Kuhn: H. Kuhn, *Helv. Chim. Acta* **1948**, *31,* 91; **1949**, *32,* 2247; **1951**, *34,* 1308, 2371; H. Kuhn, and W. Huber, *ibid.* **1952**, *35,* 1155; The didactic potential of the model has been nicely demonstrated in: H.-D. Försterling, and H. Kuhn, *Moleküle und Molekülanhäufungen: Eine Einführung in die physikalische Chemie,* Springer-Verlag, Berlin, 1983.

40) R. B. Woodward, and R. Hoffmann, *J. Am. Chem. Soc.* **1965**, *87,* 395, 2511; R. Hoffmann, and R. B. Woodward, *ibid.* **1965**, *87,* 2046; R. B. Woodward, and R. Hoffmann, *Angew. Chem.* **1969**, *81,* 797; *Angew. Chem., Int. Edit.* **1969**, *8,* 781.

SOME RECOMMENDED BOOKS

Budden, F. J., *The Fascination of Groups,* Cambridge University Press, Cambridge, 1972.
Deals with the mathematics of groups, with illustrations and examples from many fields including geometry, music, patterns, campanology, and mathematical games.

Coxeter, H. S. M., *Introduction to Geometry,* John Wiley & Sons, Inc., New York, London, 1961.
A lively, yet rigorous introduction to the subject with a strong emphasis on symmetry as the unifying thread.

Hargittai, I., and Hargittai, M., *Symmetry Through the Eyes of a Chemist,* VCH Verlagsgesellschaft mbH., Weinheim, 1986.
Covers a variety of symmetry applications at different levels of complexity; contains the most complete collection of references we know of.

Hargittai, I., Ed., *Symmetry, Unifying Human Understanding,* Pergamon Press, New York, 1986. (Also published as a special issue of the journal *Computers and Mathematics with Applications,* Volume *12B,* Numbers 1–4.)
A collection of essays by many illustrious contributors on various aspects of symmetry, from fractals, through court dances, to crystallography, music, and literature.

Jaeger, F. M., *Lectures on the Principle of Symmetry and its Applications in All Natural Sciences,* Elsevier, Amsterdam, 2nd augmented edition, 1920.
An out-of-print classic, difficult to obtain.

Jones, O., *The Grammar of Ornaments,* 1856, reprinted as *Grammatik der Ornamente,* Greno Verlagsgesellschaft, Nördlingen, 1987.
A vast collection of periodic ornaments from different periods and cultures.

Lockwood, E. H., and Macmillan, R. H., *Geometric Symmetry,* Cambridge University Press, Cambridge, 1978.
A comprehensive, largely descriptive account of symmetry for readers lacking detailed mathematical knowledge.

Loeb, A. L., *Color and Symmetry,* Wiley-Interscience, New York, 1971.
Fairly formal treatment of the symmetry of colored patterns in the plane.

Macgillavry, C. H., *Symmetry Aspects of M. C. Escher's Periodic Drawings,* A. Oosthoek's Uitgeversmaatschappij N.V., Utrecht, 1965.
An easily understandable introduction to the subject.

Rosen, J., *Symmetry Discovered, Concepts and Applications in Nature and Science,* Cambridge University Press, Cambridge, 1975.
An easy introduction into the basic concepts and terminology of symmetry. A useful source of further references.

Schattschneider, D., *Visions of Symmetry. The Notebooks, Periodic Drawings, and Related Work of M. C. Escher,* W. H. Freeman & Comp., New York, 1990.
In our opinion the best of the many books on Escher.

Shubnikov, A. V., and Koptsik, V. A., *Symmetry in Science and Art,* Plenum Press, New York and London, 1974.
Emphasis on crystallography; a special feature is the development of groups of generalized symmetry, antisymmetry, and color symmetry.

Weyl, H., *Symmetry,* Princeton University Press, Princeton NJ, 1952.
A classic essay by the great mathematician on aspects of symmetry in life, art, and scince.

Wille, R., Ed., *Symmetrie in Geistes- und Naturwissenschaft,* Springer-Verlag, Berlin, 1988.
Lectures, many in English, given at the *Symmetrie-Symposium* at the Technische Hochschule Darmstadt in 1986. Covers various aspects of symmetry, from literature, through fractals, to physics.

Williams, L. P., *Album of Science; The Nineteenth Century,* Charles Scribner's Sons, New York, 1978.
Contains reproductions of illustrations relevant to the development of 19th century science.